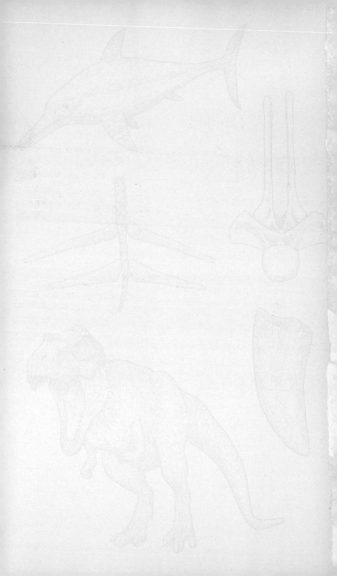

THE LITTLE BOOK OF
DINOSAURS

THE LITTLE BOOK OF
DINOSAURS

With color illustrations by Tugce Okay

RHYS CHARLES

PRINCETON UNIVERSITY PRESS
PRINCETON AND OXFORD

Published in 2024 by Princeton University Press
41 William Street, Princeton, New Jersey 08540
99 Banbury Road, Oxford OX2 6JX
press.princeton.edu

Copyright © 2024 by UniPress Books Limited
www.unipressbooks.com

Library of Congress Control Number 2024930934
ISBN 978-0-691-25989-5
Ebook ISBN 978-0-691-26013-6

Typeset in Calluna and Futura PT

Printed and bound in China
1 3 5 7 9 10 8 6 4 2

British Library Cataloging-in-Publication Data is available

This book was conceived, designed, and produced by UniPress Books Limited

Publisher: Jason Hook
Managing editor: Slav Todorov
Creative director: Alex Coco
Project development and management: Ruth Patrick
Design and art direction: Lindsey Johns
Copy editor: Caroline West
Proofreader: Robin Pridy
Color illustrations: Tugce Okay
Line illustrations: Ian Durneen

IMAGE CREDITS:
Alamy Stock Photo: 21 Heather Angel/Natural Visions; 27 Walter Myers/
Stocktrek Images; 43 Paul Fearn; 77 Mark Turner; 95 Roger Harris/Science Photo
Library; 107 Dinodia Photos; 139 Blue Robin Collectables. **Shutterstock**: 22, 58,
81, 114 Dotted Yeti; 51, 103 Herschel Hoffmeyer; 52 Ton Bangkeaw; 93 Harry
Collins Photography; 109 kamomeen; 119 Krasula; 125 Tricia Daniel;
145 Catmando. **Other**: 35 James Glazier; 133 Ian Wright; 143 PaleoNeolitic;
146 Mariolanzas. **Additional illustration references**: 15 Sam/Olai Ose/
Skjaervoy; 45 Robert T. Bakker; 47 Bob Nicholls; 79 Fred Wierum; 136 Nobu
Tamura; 147t PaleoGeekSquared; 151 Jaime Headden.

Also available in this series:

THE LITTLE BOOK OF
BEETLES

THE LITTLE BOOK OF
BUTTERFLIES

THE LITTLE BOOK OF
FUNGI

THE LITTLE BOOK OF
SPIDERS

THE LITTLE BOOK OF
TREES

THE LITTLE BOOK OF
WEATHER

THE LITTLE BOOK OF
WHALES

CONTENTS

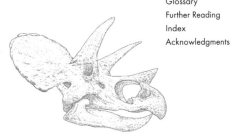

INTRODUCTION

It's a known fact that all children go through a "dinosaur phase," during which they become utterly obsessed with creatures that inhabited the Earth millions of years before humans. The great myth of adulthood is that this phase ever truly ends. No matter how old you are, it's impossible not to be captivated by this incredible group of animals.

THE PALEO PATH

Like all paleontologists, I can't remember a time when the study of the ancient world wasn't my main career goal, and inspiration to follow that path can be found all around. Museums filled with huge skeletons allow you to get a glimpse of the sheer scale the dinosaurs could reach, with individual bones that not only dwarfed me as a child but all the surrounding adults, too. Not to mention the fact that anybody, regardless of prior training, can make a groundbreaking discovery in paleontology simply by finding rocks on the beach.

Not every fossil find will make headlines, but all are undeniably magical. There really is no comparison for the rush you feel when spotting that tell-tale sign of a fossil among the pebbles, using a geological hammer to reveal it to the world for the first time in over 100 million years. And when it comes to finding fossils, surely the gold standard is to find a piece of a dinosaur.

My own journey led me to study for my Masters in Paleobiology & Evolution at the University of Bristol, where I would later return to work as the lead of the Bristol Dinosaur Project. In my job I've been privileged enough to help bring the dinosaurs back to life, not in a literal sense, but rather by sharing the latest scientific breakthroughs and even looking after more than a few giant animatronics of some species.

ABOUT THIS BOOK

This book will take a tour through the dinosaur family tree, exploring how they specialized for hunting, defending themselves, competing, socializing, and even flying. Other chapters will explore the influence dinosaurs have had on our own world, from myths and the history of science to their positions as a powerhouse of pop culture.

Dinosaurs never fail to surprise with their incredible diversity, coming in all different shapes and sizes, and modern discoveries revealing so many things we once thought impossible to know. Every day new science is published revealing more about the lost world of dinosaurs: tantalizing glimpses into how they looked, lived, and interacted together.

Our view of the dinosaur world is evolving so rapidly that if you haven't touched a dinosaur book since that childhood "phase" of wonder, you will be astonished to read how much more we know now than we did then. And even if you read a dinosaur book just last year, the exact same thing can still be said.

Rhys Charles

THE DINOSAUR CLADE

Dinosaurs were the dominant group of animals on Earth for over 160 million years, meaning they existed on the planet over 500 times as long as modern humans. During their reign they evolved into a massive array of forms, including some of the largest animals ever to have existed. This incredible group of reptiles evolved to take advantage of every opportunity in their environment. One group even took to the skies and evolved to become one of the most successful vertebrate groups today, the birds.

Currently, there are around a thousand known species of dinosaur, identified by their remains in the fossil record. There were certainly many more than that, too, with countless new examples waiting to be found.

Advances in paleontology over the last few decades have expanded our understanding of dinosaurs enormously: how they looked, lived, and died. However, their high status in pop culture means they remain a group plagued by misconceptions.

← A statue of *Megalosaurus* by Benjamin Waterhouse Hawkins featured in London's Crystal Palace Park display of 1854. It's one of the first attempts at reconstructing a dinosaur in life.

← This 1987 depiction of *Megalosaurus* by Mathew Kalmenoff shows just how much our understanding of dinosaurs has evolved over a relatively short period of time

← This is a modern reconstruction of *Megalosaurus bucklandii*, the first dinosaur to be discovered and formally named using the modern taxonomic system. This mid-Jurassic theropod stood 10 ft (3 m) tall and would have been a top predator in ancient Europe.

MAIN CHARACTERISTICS

W hen asked to define a dinosaur, most people would probably say something along the lines of "a group of large reptiles, some of which could get to enormous sizes." This definition is true enough, but it isn't perfect, since it can also be applied to many other groups of extinct animals. When classifying what makes a group scientifically unique, we need a much more precise set of characteristics.

ARCHOSAURIA

Dinosaurs are part of the reptile group Archosauria, which today is represented by two living groups of animals: the crocodilians and the birds. On the evolutionary tree, dinosaurs effectively fall in the space between these two groups, their last common ancestor with crocodiles living about 240 million years ago (Mya) and birds having originated within dinosaurs themselves.

The skulls of dinosaurs feature two additional fenestra (openings), as well as the orbit and nasal openings. This puts them in a group of reptiles called diapsids, separate from the synapsids (such as mammals), which have one such fenestra, and the anapsids (like tortoises) which have none.

BLOOD TEMPERATURE

Today, crocodilians are cold-blooded (ectothermic), while birds are warm-blooded (endothermic). Falling in the middle of these groups, it has long been debated when endothermy evolved in dinosaurs. Evidence from fossil biomolecular data (looking for the "waste products" of high metabolism) suggests that endothermy evolved before the base of the dinosaur family tree, meaning that by default dinosaurs were warm-blooded. However, later, large ornithischians, such as *Stegosaurus*, may have lost this feature and switched to ectothermy.

LEG POSITION

One of the main defining features of dinosaurs is that they held their legs directly underneath their bodies. This differs from crocodilians which have a more sprawling gait, their legs held off to the sides. Holding their legs directly underneath the body in this way is what allowed dinosaurs to reach such large sizes, as they were better able to support their weight without putting massive strain on their joints. It also allowed for quite energy-efficient running as bipeds, making the early dinosaurs more nimble than the other archosaurs with which they shared their environment.

HIP DIVISION

Very early on in their evolutionary history the dinosaurs split along two distinct evolutionary paths to form two groups: the saurischians and the ornithischians. All dinosaurs fall into one of these two groups, which is determined by the structure of their hips. Hips are made of three bones: the ilium, pubis, and ischium. In the saurischians, or "lizard-hipped" dinosaurs, the pubis bone runs down and slightly forward, toward the animal's head. It also regularly ends with a flattened keel or boot structure. In the ornithischians, or "bird-hipped" dinosaurs, the pubis is reversed. Instead of running down and forward, it runs back toward the tail, usually lying parallel to the ischium. This divide happened right at the base of the dinosaur family tree, with the early saurischians evolving into the theropods (i.e., *Tyrannosaurus rex*) and sauropods (i.e., *Diplodocus*), and the ornithischians into the ornithopods (i.e., *Hadrosaurus*), marginocephalians (i.e., *Triceratops*), and thyreophorans (i.e., *Stegosaurus*). Counterintuitively, it was the "lizard-hipped" dinosaurs that evolved into birds, rather than the "bird-hipped" dinosaurs.

→ Dinosaur hip structures are saurischian (left) or ornithischian (right). The top bone is the ilium, and the pubis (on the left) and ischium (right) are beneath.

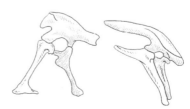

FOSSILS

Our primary evidence for dinosaurs comes from their fossilized remains, dug out of rock millions of years after the animals lived. Dinosaur fossils were formed when the bodies of dinosaurs became buried in sediment and the organic remains decayed and were replaced by mineralized rock. The original material of the dinosaur is long gone and only an abiotic record is left.

Unfortunately, only the tougher sections of the body were favored for preservation, which is why most finds are purely skeletal and lack soft body elements. Even then, only a tiny percentage of the living animals had a chance of becoming a fossil and then later found. A lot can interfere with the process across millions of years.

Sometimes, when conditions are perfect (fast burial, low oxygen, little disturbance of the body), dinosaurs can be exceptionally well preserved, retaining details such as skin, feathers, and even stomach contents. The sites where these fossils are found are called *Lagerstätten* (German for "storage-places").

↓ Ammonites and other marine invertebrates are the most common fossils because their resistant shells and lifestyle meant they could be buried quickly and easily by sediment.

↓ Teeth are numerous and durable features, making them common fossils. The shape of teeth, like this *Iguanodon's*, can provide information about the animal's diet.

→ It is exceedingly rare to find articulated skeletons preserved in the rock. It's far more likely that only fragments will be recovered, making reconstructions a challenge. However, in some circumstances, as in this *Sinosauropteryx* specimen, even the finest details are preserved.

THE FIRST DINOSAURS

T he very first dinosaurs were bipedal carnivores. Drawing exact lines is difficult in paleontology, so there is no definitive answer when trying to name the first true dinosaur. However, several genera that appeared in the Triassic Period (252–201 Mya) and showed very primitive features can help us understand what those first dinosaurs were like.

NYASASAURUS

The earliest known fossil confirmed to be from the dinosaur group is *Nyasasaurus*. Found in Tanzania in 1932, this dinosaur is dated to about 233 Mya, in the middle of the Triassic. Whether or not *Nyasasaurus* is truly a dinosaur is still debated, largely due to the fact that it is only known from partial remains. The structure of the humerus is the strongest evidence for inclusion, as the shape of a ridge for muscle anchorage on the bone is one that is only seen in dinosaurs.

HERRERASAURUS

A more complete early dinosaur is *Herrerasaurus*. Dated to about 230 Mya, this bipedal carnivore could be mistaken for a theropod thanks to its skull and stance. The structure of the hip gives it away as an earlier dinosaur form, lacking as it does certain characteristics that would only evolve later. *Herrerasaurus* is now viewed as a very early saurischian dinosaur, a precursor to the theropods and sauropods (see Chapters 4 and 5). It stands as a testament to how quickly the dinosaurs diversified at the end of the Triassic into all the forms we know today.

FEATHERED BY DEFAULT?

Feathers had been known in bird-line dinosaurs for a long time, and it was presumed that they evolved here in the theropod lineage, millions of years into dinosaur evolution. However, the discovery of dinosaurs like *Kulindadromeus* put this into question. *Kulindadromeus* was a small ornithischian dinosaur from the Jurassic Period (201–145 Mya) that clearly showed feather-like structures. If present in this branch of the dinosaur tree, entirely separate to the birds, it could mean that either feathers evolved multiple times in dinosaurs or only once at the very base of the tree, and the ancestor to all dinosaurs potentially possessed feathers already. Feathers and quill structures have also been discovered in some ceratopsians (the group of sometimes horned ornithischian dinosaurs, including *Triceratops*), lending further evidence to this idea.

↘ *Herrerasaurus* was first discovered in South America in the late 1950s. As one of the oldest known dinosaurs, its exact place in the family tree is debated.

WHAT ISN'T A DINOSAUR?

One issue with the huge popularity of dinosaurs is that the term often includes various imposters—creatures that are regularly called dinosaurs, which aren't. Some of these are simply other reptiles who shared their world, but others are separated from the dinosaurs by millions of years.

DIMETRODON

The most typical of these imposter species is the sail-backed *Dimetrodon*, an animal that existed 50 million years before the first dinosaurs. Far removed on the evolutionary tree, *Dimetrodon* is a synapsid, a creature closer to the early mammals than the dinosaurs. Although not as sprawling as a lizard, the legs of *Dimetrodon* were also not held as pillars under the body, as is seen in dinosaurs. The jaw joint also differs from that of dinosaurs, showing a ridge on the articular bone (found at the lower jaw joint) that appears as a precursor to the middle ear bones of true mammals millions of years later. An apex predator of its time, the sail was most likely used as a display for mates and for intimidating rivals.

ICHTHYOSAURS

The marine reptiles that swam the seas during the dinosaur age are also regularly grouped among them. This is particularly unfair considering what an amazing group they were in their own right.

Ichthyosaurs were marine reptiles with distinctly fish-shaped bodies, often complete with a protruding dorsal fin and shark-like caudal, or tail, fin. Their jaws were extended into long rostrums (snouts), and many had large eyes for hunting effectively in water. First evolving at a similar time to dinosaurs, the ichthyosaurs existed alongside them for over 130 million years. In that time, they evolved into a myriad of different niches, from the giant predatory *Temnodontosaurus*, to sediment-sifting specialists like *Excalibosaurus*. One relative, *Hupehsuchus*, was even a filter feeder.

SAUROPTERYGIANS

Not all marine reptiles were fish-shaped. The sauropterygians evolved a unique way of moving through the water, using four undulating flippers. Like ichthyosaurs and dinosaurs, they, too, were large reptiles and lived alongside both groups. Initially, sauropterygians contained the durophagus (shell-crushing) placodonts, some of which had armored carapace shells resembling those of sea turtles. However, the placodonts went extinct at the end of the Triassic.

Best known of the Sauropterygia were the plesiosaurs. The necks of these fish-eating reptiles (such as *Elasmosaurus*) could reach over 23 ft (7 m) in length and their tails were highly reduced. Their body plan is familiar, since it formed the basis of the Loch Ness Monster myth.

Whereas plesiosaurs had relatively small heads on long necks, the pliosaurs had the opposite. The jaws of these top marine predators could measure nearly 8 ft (2.5 m)—nearly one-third their total body length)—and were full of huge conical teeth that evolved for hunting other large animals.

MOSASAURS

The mosasaurs were a group of marine reptiles that evolved in the Cretaceous Period (145–66 Mya). Not dinosaurs, they were also not related to the ichthyosaurs or sauroptergians. Instead, mosasaurs were part of the Squamata order, a group that also includes snakes and lizards. Taking the niche occupied by the pliosaurs as they went extinct, the mosasaurs were large predators, growing up to around 33 ft (10 m) in length. Although they had fin-like limbs, they moved by undulating their body like a fish, propelling themselves with a paddle-like tail. Eventually, they would go extinct with the dinosaurs, leaving the sharks as the top marine predators.

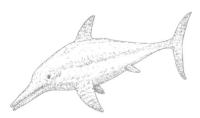

← Ichthyosaurs had a fish-like shape, which was the result of convergent evolution, whereby animals evolve similar solutions when they are faced with the same environmental challenges.

IN THE AIR

For 160 million years alongside the dinosaurs, the dominant group of flying vertebrates were the pterosaurs. Although not actually dinosaurs themselves, they are quite closely related and often called a "sister taxon," having likely diverged from a common ancestor in the Early Triassic.

WING FINGER

Pterosaurs did not have any flight feathers and their wings were built very differently to those of birds. The structure of the wing was supported by one enormous and hugely modified third finger. A thin membrane of skin stretched from the tip of this finger and anchored at the ankle, creating a massive surface area for generating lift. This was increased by the propatagium, a membrane that covered the elbow of the wing.

The design of the wing was extraordinarily effective, allowing pterosaurs to evolve into the largest animals ever to fly. The largest of them all, the azhdarchids of the Cretaceous, had wingspans of over 33 ft (10 m). An individual finger bone from one of them, *Quetzalcoatlus*, could be over 22 in (56 cm) long. For comparison, the largest flying bird alive today, the Wandering Albatross (*Diomedea exulans*), has a wingspan of 9.8 ft (3 m).

WEIGHT SAVING

Everything about pterosaurs was geared toward being more efficient fliers. Early pterosaurs had jaws with teeth, but by the Cretaceous these had been replaced with lightweight beaks. Similarly, early small species had long, bony tails for air maneuverability, but these were lost as they evolved to become larger.

The entire skeletons of pterosaurs were hollow, filled with air sacs to make them far lighter than their size would suggest. Internal structures provided strength in life, but, unfortunately, they are very fragile in death, easily lost and damaged by the fossilization process, which makes pterosaur material extremely rare.

PTEROSAUR CRESTS

Several pterosaur species are known to have had elaborate structures on their heads, the most recognizable being the large, posterior-facing ridge of bone in *Pteranodon*. This is the typical image of a pterosaur, although others could be stranger, such as the antler-like crests of *Nyctosaurus*. These structures not only worked for display, but also in flight. They might have made the animals look top-heavy, but it meant their center of gravity was shifted forward, making their launches more effective and balancing their flight.

↓ A specimen of *Pterodactylus*, whose name means "Wing finger." This animal is thought to be behind the popularity of the word pterodactyl, despite no pterosaur genus actually bearing that name.

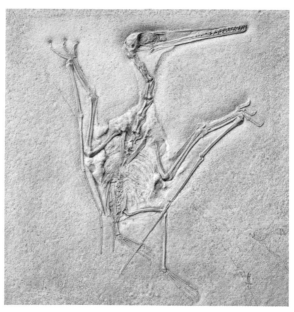

BEFORE THE DINOSAURS

Dinosaurs are so linked to the notion of prehistory that many find it hard to imagine a time before them. But life history did not start with the dinosaurs. In fact, by the time they first appeared, life had already been evolving for over 3 billion years. Life began in the Earth's waters as single-celled organisms, taking billions of years to evolve into truly complex forms.

The Cambrian Explosion of about 500 Mya is often signposted as being the major trigger in the rapid diversification of complex life, and indeed it is to here that many of the familiar branches of the tree of life can trace their beginning, including primitive vertebrates: simple jawless fish with basic internal notochords that had not yet developed into what we would call a spine.

CONQUEST OF THE LAND

Several million years later, arthropods and plants made the exploratory journey to colonize dry land, while fish with specialized lobe-finned limbs made a similar venture. All tetrapod life, including dinosaurs and ourselves, evolved from these fish, which is why we share such similar bone and limb structures.

By 350 Mya, life had truly conquered the terrestrial landscape. Much of the world was covered in tropical rainforest in a time named the Carboniferous because this is the period from which much of our carbon-loaded coal fuel originates. The giant invertebrates, like the millipede *Arthropleura*, which measured 8 ft (2.5 m), and the hawk-sized dragonfly relative *Meganeura*, are the more popular stars of the period, but it was also when amphibians first evolved hard-shelled eggs and thicker scaled skin. These innovations allowed them to move away from the water. The reptiles had arrived, and theirs was a dynasty that would dominate the Earth for the next 200 million years.

THE PERMIAN

By the Permian Period (299–252 Mya) there were two dominant groups of reptiles: the archosaurs and the therapsids. The archosaurs included relatives of today's crocodiles, although they were much more diverse. Not just water-based predators, they even included bipedal herbivorous forms. The therapsids were the mammal-like reptiles, ranging from small, burrowing cynodonts to the large, saber-toothed, predatory gorgonopsids. These creatures had yet to evolve the distinctive fur that would mark them out as mammals, but their jaw structure and the shapes of their skulls give away their lineage.

← Complex eyes were among the main innovations of the Cambrian Explosion, allowing for the evolution of huge arthropod predators like anomalocarids.

END-PERMIAN EXTINCTION

At the end of the Permian Period (250 million years ago), the planet entered a crisis. Extreme volcanic activity across what is now Siberia plunged the climate into chaos, warming the world through the emission of massive levels of carbon dioxide, turning oceans acidic, and starving the waters of oxygen.

The effects on life were dramatic, with some marine areas suffering a loss of over 90 percent of their biodiversity. Complex life had suffered through two mass extinction events before, but this third one was the biggest it had ever faced. Victims included some icons of paleontology, like the trilobites, and whole swathes of the large reptile groups that had dominated the planet for a 100 million years.

This extinction event had effectively wiped the slate clean and created a new world, waiting to be claimed. The race was on and the stage set for the entrance of the dinosaurs.

← Coral diversity was heavily impacted by the End-Permian extinction, with forms such as horn-shaped rugose corals disappearing forever.

↙ The trilobites were perhaps the most famous victims of the End-Permian extinction. Icons of paleontology, they thrived for over 250 million years, but went extinct 20 million years before the first dinosaurs had even evolved.

↓ The saber-toothed gorgonopsids were the top predators prior to the End-Permian extinction, but they and their relatives would not survive through to the Mesozoic Era. Belonging to a group called the therapsids, these impressive beasts were part of the same animal clade from which mammals would eventually emerge.

THE TRIASSIC PERIOD

What is colloquially called the time of the dinosaurs is a geological era known as the Mesozoic. This expanse of 186 million years (252–66 Mya) is further split into three periods, the first of which is the Triassic (the others being the Jurassic and Cretaceous, which are covered later in the chapter). A common misconception is that dinosaurs ruled the planet from the beginning of this era, springing into the empty niches left behind by the End-Permian extinction. However, it wasn't until much later in the Triassic that the dinosaurs rose to prominence.

PERMIAN RECOVERY

In the aftermath of the End-Permian extinction, the world was still quite a desolate and brutal place. Deserts were the dominant habitat, covering vast areas of land. A "hot house" earth system meant that carbon dioxide levels were potentially up to three times higher than they are today.

Biodiversity could be so low that in some places over half of all the vertebrate fossils found may belong to a single genus, a charmingly squat dicynodont called *Lystrosaurus*. The name may sound dinosaurian but this creature was actually a member of the mammal-like therapsids.

THE CARNIAN-PLUVIAL EVENT

In the middle of the Triassic came an enormous change from an arid climate to one of extreme humidity and rainfall. Evidence for this is found in rocks across the planet, with volcanic activity in North America thought to have been a trigger. Unlike most other similar events, however, this time is not remembered for the animals lost, but instead for those that succeeded in the aftermath. Following the Carnian-Pluvial Event, the fossil record shows an explosion in dinosaur diversity and dominance. In essence, this period of around 2 million years of climatic upheaval created the ecosystem blueprint for the rest of the Mesozoic.

CONTINENTAL
POSITIONS

The Triassic was a time of "super" systems. All the major landmasses on Earth were joined together to form the supercontinent, Pangaea. In total, Pangaea measured approximately 57.5 million square miles (149 million square kilometers). Surrounding Pangaea was a super-ocean called Panthalassa, which covered 70 percent of the globe. However, it wasn't the only ocean, as the Tethys and Paleo-Tethys could be found on the eastern side of Pangaea, separating it from the only other extensive land systems, a series of islands called Cimmeria that now make up parts of southern Asia.

TRIASSIC EXTINCTIONS

A few million years into their reign and the dinosaurs would be faced with the first major test of their rule as the dominant life form on the planet. The end of the Triassic Period was marked 201 Mya by a cataclysmic event: another mass extinction. Long-term fluctuations in both climate and sea level, perhaps caused by increased volcanic activity in central Pangaea as the continents began to break apart, is thought to have driven this extinction. It was felt hardest in the sea, particularly among the invertebrates, which struggled to survive in acidified oceans.

The dinosaurs made it through this collapse at the expense of many of their fellow amazing archosaurs, such as the armored aetosaurs, and other terrestrial crocodile relatives. The crocodilians would never recover their former diversity. From the point of view of the dinosaurs, it was just another loss to the competition, and their grip on the throne of the land grew tighter.

↓ *Prestosuchus* was a terrestrial pseudosuchian from the Triassic Period. A terrestrial predator hunting other large reptiles, its skull is very different to those of its closest living relatives, the crocodiles.

↘ The skulls of modern crocodiles are flattened in an adaptation to their aquatic environment, showing a much-reduced diversity compared to other pseudosuchian relatives in the Triassic Period.

→ The backs of aetosaurs were covered in protective armor and long spines for both ornamentation and defense. Existing for around 30 million years, they are part of a group called pseudosuchians, which are distantly related to the group that would become modern crocodiles.

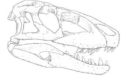

THE JURASSIC PERIOD

S tarting a little over 200 Mya, the Jurassic was defined by the breakup of Pangaea and shifting tectonic plates. A major rift separated the northern landmass, known as Laurasia, from that to the south, Gondwana. These two supercontinents would make up much of the dinosaur world, although even they would start to break apart by the end of the Jurassic. India and Africa were separated from Laurasia by the Tethys Ocean, the Atlantic still only just beginning to grow.

MARINE INSTABILITY

In the Jurassic Period, oceans cycled through periods of mass anoxia (a lack of oxygen dissolved in the water), which led to several extinction events, although not quite on the same scale as the big mass extinctions. The largest, the Toarcian Ocean Anoxic Event, was likely triggered by volcanic activity in southern Africa, 183 Mya. Increased carbon dioxide levels in the atmosphere and the disruption of ocean currents caused widespread anoxia, devastating marine invertebrates. The process also acidified the oceans, making it harder for creatures to create stable calcium shells, something that we also see affecting today's marine life.

A WARMER CLIMATE

Thanks to elevated carbon dioxide levels, the temperature of the Jurassic was warmer than it is today. Some environments, like rainforests, were rare, while temperate forests and scrubland thrived. At its warmest it's thought the climate may have reached a global average 14.4°F (8°C) higher than today. Temperatures did dip at the extremes of the planet, and cycled over the years, although they remained warm enough that no permanent ice caps were able to form at the poles. Even Antarctica was free from ice and home to many species of dinosaurs that thrived in forests there.

The lack of any ice caps at the poles meant much higher sea levels across the globe. So much of what we know about the Jurassic comes from shallow sea deposits that span large sections of what is today the dry land of Europe. Although many large landmasses remained, much of the low-lying lands were a series of islands. North America, too, was dominated by water, with the west of the continent split from the east by a huge inland sea called the Western Interior Seaway. On today's map this sea would stretch from Utah to Iowa.

FLORA AND FAUNA

By the Jurassic most of today's major animal groups had established themselves. Obviously dinosaurs were a novelty, but alongside them were familiar reptiles such as true lizards and legless forms of proto-snakes, which did not exist until the very end of this period.

It is also in the Jurassic that the fossil record shows definitive mammals first appearing. They remained small but had begun evolving into the main groups we see today, including the eutherians, the group that produced live young (to which we belong). Another major group to appear first in the Jurassic were the birds, although this will be covered in more detail in Chapter 7.

The plant world was dominated by ferns, cycads, and monkey puzzle trees (*Araucaria*), all of which are still present today. They are not as dominant now as they were in the Jurassic when they did not face competition from other major plant groups yet to evolve.

→ The giant tree ferns that dominated the Jurassic still grow today, although they are less dominant. Some species in Australia and New Zealand can reach almost 65 ft (20 m).

THE CRETACEOUS PERIOD

T he final period of the Mesozoic Era, the Cretaceous (145–66 Mya), was enormously important in Earth's history. A time of multiple revolutions (periods of rapid evolution in many lineages), this was also when dinosaur diversity reached its peak. As with the other periods of the Mesozoic, the climate was warmer than it is today, with no permanent ice caps at the poles. But on average, the Cretaceous was cooler than both the Triassic and Jurassic.

GETTING INTO POSITION

A map of the Earth in the Cretaceous would be far more familiar to us, with some continents closer to their modern positions. However, significant differences remained. India was still adrift in the middle of the ocean, about 20 million years away from colliding with mainland Asia. Due to a lack of data, it's difficult to tell whether Australia had yet separated from Antarctica, breaking the final remnant of Gondwana.

Sea levels remained high, with much of northern Africa and central Asia under water, but some of the inland seas over Europe and North America had begun to recede massively, opening up new fertile habitats in the Northern Hemisphere.

THE RISE OF FLOWERS

One of the most important evolutionary radiations in history largely took place in the Cretaceous, although its effects would extend all the way to the modern day. The Angiosperm Terrestrial Revolution describes a period of time in which flowering plants first evolved and began to dominate land-based ecosystems all over the world.

Angiosperms (flowering plants) are nearly ubiquitous on Earth today, making up 90 percent of all plant life. They support a whole network of other organisms, including thousands of species of pollinating insects. The arrival of flowering plants led to an explosive adaptive radiation of animals to explore this new niche. The knock-on effects of this were literally world-changing. Not since the first trees, had plants made such an important evolutionary leap. How this

← Fossil evidence and the study of the modern flower family trees allow for recreations of how some of these very earliest flowers might have looked.

affected the dinosaurs we will look at specifically later in this book (see Chapter 8, page 102). Although this revolution had potentially begun earlier than the Cretaceous, and certainly extended beyond it, the Cretaceous was probably the period in which the most significant changes occurred.

SEEDING CHANGE

Flowers weren't alone, as the Cretaceous also sees fossil evidence of some of the first grasses, another plant we now can't imagine the world without. However, modern forms of grass, and their subsequent dominance of open landscapes, were still far away. Grassland habitats as we know them today would not evolve until after the dinosaur age.

MARINE REVOLUTIONS

In the seas, another revolution that began in the Jurassic was coming to fruition. The Mesozoic Marine Revolution saw an increase in the complexity of hard-shelled invertebrates, in what appears to have been a response in an arms race between the invertebrates and durophagus fish (these had hard, round teeth specially adapted for crushing shells). The results were widespread, but perhaps most notable was the appearance of the brachyurans (true crabs), their compact bodies allowing them to hunker down and avoid predation among rocks.

Marine reptiles were also seeing a shift. The ichthyosaurs and plesiosaurs were declining, replaced in many places by mosasaurs, relatives of lizards that had taken to the seas and evolved to sizes of over 35 ft (10.5 m).

MYTHS AND FOLKLORE

Dinosaur fossils are found all over the world, often by people who aren't even looking for them but who accidentally stumble across gigantic bones emerging from the rock. So it is hardly surprising that humans have been finding—and becoming fascinated by—fossils since long before we attempted to describe them scientifically.

EARLY INTERPRETATIONS

With no real context as to what they might have been, it is difficult to know what the people of ancient civilizations would have made of dinosaur fossils, but there are pieces of evidence to suggest that they may have been the basis of mythical stories passed down the generations through oral history. The most natural, although still very speculative theory, is that dinosaur bones might go some way to explaining the prevalence of dragon stories in various cultures across the globe.

Of all early interpretations of fossil discoveries, it is perhaps the Native American people whose explanations came closest to the truth. Although they may not have accurately recreated the looks or relationships between the creatures, their stories referred to them as the remains of animals that roamed the Earth many eons before the arrival of humans, giving them names and respecting their lost lives.

~ Practical purposes ~

Other fossils found more practical uses, with fossilized shark teeth used as decorative jewelry as early as 4000 BC. It's thought that the teeth of the famed giant shark Megalodon were even used as tools to create grooves in ceramic works in Malta in about 1500 BC.

INSPIRING ART

It has been theorizied that the mythological winged griffins could, in part, have been based on *Protoceratops*, the skeletons of which have frequently been found scattered over some parts of the Gobi Desert. They certainly tick a lot of the boxes: four-legged, squat reptilians with pointed beaks and elaborate skull frills. But amid the conjecture, there is harder evidence of ancient civilizations mixing with fossils. Although it is not depicting a dinosaur, it's very hard to argue that the "Monster of Troy" vase of around 550 BC (shown above) does not depict the giant fossilized skull of some ancient creature, which the Greeks likened to a vanquished former monster of the land, in many ways accurately.

TERRIBLE LIZARDS

By the start of the 19th century, natural scientists had been categorizing the vast diversity of life on Earth into taxonomic classifications for many years. Animals and plants were given species names following the system pioneered by Swedish botanist Carl Linnaeus (1707–78) a century earlier. The natural world was being studied in detail like never before, and with dinosaur bones so numerous and widespread it was only a matter of time before they became the subject of attention.

THE GREAT LIZARD

It started in the United Kingdom, with one of the most iconic individual fossils ever found. From a quarry in Stonesfield, near Oxford, came a single jawbone with a dagger-like tooth standing proudly in the center. Over the next few years, other large bones followed, from creatures unlike anything known alive at the time.

Zoologist William Buckland (1784–1856) acquired many of these fossil bones and concluded that they belonged to some form of gigantic predatory reptile. In 1824 he published his work and gave the fantastical creature a memorable name, *Megalosaurus*, meaning "Great Lizard." This was the first formally named dinosaur.

~ Discovered twice ~

We now believe that *Megalosaurus* had already been discovered once before. In 1676, the end of a large saurian femur was unearthed. Based on the bone's appearance, it was hypothesized to have come from a giant human. Said to resemble a rather intimate part of the human anatomy it was given the unfortunate name of *Scrotum humanum*. Sadly, the fossil in question has since been lost and all we have to go on is an illustration.

THREE OF A KIND

The great lizard of Stonesfield wasn't an isolated incident, as the remains of more disproportioned reptiles were found in southern England. Teeth discovered by Mary-Ann Mantell (1795–1869) and Gideon Mantell (1790–1852), which looked like those of an elephant-sized Iguana lizard, were deemed to have come from *Iguanodon*. A few years later, the remains of a third creature emerged from West Sussex. Unlike the previous two, the fossils of this animal were partially articulated and seemed to have some level of protective armor. Uncovered in Tilgate Forest, it was dubbed *Hylaeosaurus*, meaning "Lizard of the Woods."

With three new species of giant reptile from the past now formally described, it was clear that the stunned scientific world would need to classify them in a collective group. It was Sir Richard Owen (1804–92), founder of London's Natural History Museum, who first proposed just that. Using the characteristics of the three named examples, he announced in 1842 that they would become the foundation of a new group of animals, known as "terrible lizards." Scientific classification uses Latin and so the translation of the term was used: *Dinosauria*.

SAURUS

It is because these fossil finds were placed in the group "terrible lizards" that the names of so many genera of dinosaur end with the suffix *-saurus*, meaning "lizard." It was a rather fateful choice, as now dinosaurs are stuck with that name and reputation even though they are not close to lizards in physiology or phylogeny. In spite of this, the naming of dinosaurs in this manner persists to this day, although it is now more out of tradition and for the sake of being recognizably a dinosaur to the public, rather than due to any scientific merit.

↘ The initial fossil discovery of *Megalosaurus* shows a fragment of the lower jaw, with a defined protruding tooth. The fossil was at first assumed to have come from an animal like a crocodile.

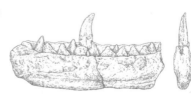

AROUND THE WORLD

Once the dinosaurs had been placed in a group, it wasn't long before people were reporting findings all over the world, and it became clear that dinosaurs were not just reserved for the south of England. Teeth found at the Judith River Formation (in Montana) were attributed to *Dinosauria* in 1854, the group's first announcement in North America. They were later said to have come from the carnivore *Daspletosaurus*, although this remains uncertain. That same year, the long-necked *Massospondylus* became the first named African dinosaur, found in South Africa. *Argyrosaurus* of Argentina was named in 1893 and *Rhoetosaurus* of Australia in 1926.

Some discoveries made years earlier were reassigned to the new group, as was the case for the first Asian dinosaur, based on a few tail vertebrae of a sauropod from Jabalpur, India, in 1828. It took until 1986 for the first Antarctic dinosaur, *Antarctopelta*, to be discovered, but with it came proof that dinosaurs were a global phenomenon.

↑ Large, mysterious vertebra bones were found in Jabalpur, in India, in 1828. But it was only many years later that it was realized these came from a dinosaur.

↑ *Daspletosaurus* teeth are typical of predatory dinosaurs; serrated on the side to cut through flesh, and with a large triangular shape.

↙ *Antarctopelta* is known only from fragments weathered by the tough conditions of James Ross Island, off the coast of Antarctica.

But more complete specimens of closely related species from South America allow for reconstructions of this armored dinosaur.

THE BONE WARS

One of the greatest periods of discovery in those early days of dinosaur paleontology owes itself to a scientific rivalry. In the late 1800s, dinosaur fossils were the new gold rush for North America, and two eminent scientists of the time led (and funded much of) the charge: Edward Drinker Cope (1840–97) and Othniel Charles Marsh (1831–99).

Immediately the challenge became for each to find more and grander dinosaurs than the other, by any means necessary. Public slander, bribery, theft of specimens, and even actively destroying each other's finds took place and became worryingly commonplace in paleontology for years. It was a war of science fueled by fragile egos. Their feud was bitter, the stories of sabotage and scandal leaving a permanent mark on the field, but its other legacy was the discovery of some of the most famed names in the dinosaur world, including *Triceratops*, *Diplodocus*, and *Stegosaurus*.

↓ Orniel Marsh's classic reconstruction of *Stegosaurus* may not be particularly accurate from a scientific point of view, but it acted as an iconic blueprint for how many people see dinosaurs some 150 years later.

→ *Allosaurus fragilis* was another famous dinosaur found during the Bone Wars. A top predator of the Morrison Formation of ancient North America, it would later gain the media nickname of the "lion of the Jurassic."

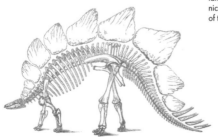

DINOSAUR CENTURY

T he huge popularity of dinosaurs, both with the general public and in scientific research, only grew at the turn of the 20th century as even more discoveries revealed just how abundant and widespread they had been as a group.

FAMED NAMES

One of the most prolific paleontologists of the early 1900s was the German Friedrich von Huene (1875–1969), who named around 30 dinosaur genera. Although many of these names were based on fragmentary evidence and have since been cast into doubt, some of his other work, such as being the first to recognize and name the "Sauropoda" as a group, has stood the test of time.

There is no shortage of interesting figures from this period of dinosaur science. They include American Barnum Brown (1875–1963), who discovered *Tyrannosaurus rex*, and the truly fascinating Hungarian Baron Franz Nopcsa (1877–1933), who pioneered work on dinosaur physiology (as well as formally nominating himself as King of Albania).

DESTROYED DINOSAURS

Despite being extinct for millions of years, dinosaurs were not untouched by the events of the 20th century. Small museum pieces were easy to move during the bombing raids of the Second World War, but this was not the case for the heavy rock remains of giant dinosaurs, many of which were left behind in the hope that they would not be hit. Unfortunately, many were unlucky. The most high profile of these were the remains of the huge theropod *Spinosaurus*, which were obliterated when Munich's *Paläontologishes* Museum was bombed in April 1944.

EGG EXPEDITIONS

One influential but controversial series of expeditions in dinosaur paleontology were those undertaken by the American Roy Chapman Andrews (1884–1960) in the 1920s. His group explored the vast bone fields of the Gobi Desert, making many remarkable discoveries, including the first verified fossils of dinosaur eggs (belonging to the beaked theropod, *Oviraptor*). However, one of these very eggs caused great controversy, as he put it up for auction upon returning to America. Andrews had assured the ruling government (China at the time) that the eggs had no monetary value, and they in turn accused him of conning them, plundering their precious finds for profit in the United States. As such, any further expeditions were banned for over a year.

THE DINOSAUR RENAISSANCE

Ever since the sensational discovery of dinosaurs and the huge sizes they could reach, it was generally assumed that any animal so large must have been a lumbering giant, slowly plodding its way across the primordial world. Surely it was impossible for such large creatures to move with any rapidity or grace.

TERRIBLE CLAW

The spark for the revolution in thinking was the discovery of a new dinosaur genus, *Deinonychus*, in 1963 (the first remains were discovered in the 1930s but never formally described). *Deinonychus* was an American raptor, a group known for many distinctly bird-like features, including the sickle claw on the toe for which the group is infamous (the genus name even means "Terrible Claw"). These features caused the paleontologists of the time to bring back a theory first proposed only a few years after the publication of Darwin's theory of evolution: that birds had evolved from dinosaurs. It's not fully known why this idea had fallen out of fashion, although some blame a popular work by Danish paleontologist Gerhard Heilmann, *The Origin of Birds* (1926), which dismissed the idea as fanciful. But now that it had been raised again it almost appeared too obvious, and with every specimen checked, more evidence mounted that the two groups were connected.

BIRD BLOOD

Unlike reptiles, birds are warm-blooded (endothermic), just like mammals. So if they evolved from dinosaurs, it is possible that some, if not all, dinosaurs were also warm-blooded, which would have enabled them to lead far more active lives. It would be some time before technology would advance enough to accurately report on these claims. Detailed studies of the bones of dinosaurs (and other prehistoric groups) using infrared spectroscopy showed the presence of tell-tale biomolecules indicative of a high metabolism, like that of birds. This suggests that endothermy may well have evolved at the base of the dinosaur tree, and the more active lifestyle proposed in the "Dinosaur Renaissance" can be applied to the whole group.

COLD-BLOODED COMPARISONS

The quick assignment of the dinosaur group to the Reptilia may well have had a hand in shaping our perception of them. After all, scientists at the time would have compared them directly to living reptiles, such as lizards and crocodiles. These groups are cold-blooded (ectothermic), meaning they are reliant on external heat sources to warm their bodies. As such, they spend much of their time basking in the sun and reserving energy by moving as little as possible until they have to, usually to feed. With dinosaurs being so closely compared to such reptiles, and because of the extra weight conferred by their enormous body sizes, an even more lethargic lifestyle was presumed for them.

INSPIRING IMAGES

What really boosted the new understanding of dinosaurs, however, was the famous image that accompanied the research. The paleontologist Robert Bakker (b.1945) reconstructed the animal as a fleet-footed and nimble hunter, drawn in the act of running and showing a lightweight and athletic build never before given to a dinosaur. This image would provide the blueprint for the way we picture dinosaurs to this day. Starting from a piece of artwork and exploding into a myriad of new paradigm-shifting research in the years that followed, this period of paleontology in the 1960s and early '70s is known as "Dinosaur Renaissance."

THE THEROPODA

T he very first dinosaurs were likely all carnivorous; light and nimble hunters, they fed mostly on insects and other small reptiles, prey they were perfectly adapted to handle with their razor-sharp teeth. While some lineages went on to evolve herbivory and take advantage of the plentiful vegetation, one group stuck to their meaty diets and evolved into an amazing array of effective killers. These were the theropod dinosaurs.

MORPHOLOGY

Many of the features retained by the dinosaurs through the whole Mesozoic Era (252–66 Mya) can be seen in those Triassic pioneers. The theropods were saurischian dinosaurs (with the lizard-like arrangement of hip bones) and they were bipedal, leaving their arms free for use in catching prey, meaning that many evolved particularly sharp claws. The earliest members of the group possessed up to five digits on each hand, although this reduced to three in most later theropods.

EXTINCTION SURVIVORS

We do know that despite the challenges, the theropods survived through the end-Triassic extinction. Finds from around this time are rare, but specimens from shortly after the extinction, such as *Dracoraptor* of Wales, give some clues as to how the group survived. The small serrated teeth and advanced sense of smell possessed by *Dracoraptor* hint that it was likely a specialist in small prey and had a scavenging lifestyle. The dinosaur was lean and nimble, with a body size a little over $6^{1}/_{2}$ ft (2 m) in length. Being a fast opportunist during tough times helped the dinosaurs survive where the other top reptile carnivores failed.

GHOST RANCH

Much of what we know about the Triassic theropods comes from one location, Ghost Ranch in New Mexico, where it's estimated over a thousand individuals of *Coelophysis* have been discovered. These dinosaurs, a little over 3 ft (1 m) in height, had fairly narrow skulls atop relatively long, curved necks. Their lightweight jaws suggest that though they were diversifying, during the Triassic the theropods were sticking to small prey items. It was once thought that small bones in the body cavity of one *Coelophysis* were evidence that this dinosaur may have been involved in cannibalism, perhaps an extreme survival tactic during the harsh Triassic environment. However, we now know that the remains belonged to a different species of small reptile.

↘ Found in rocks from the earliest Jurassic of Penarth, Wales, *Dracoraptor hanigani* shows a fairly typical body plan for early theropods.

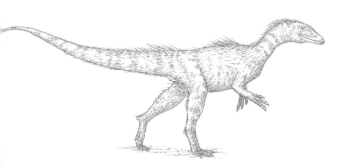

JURASSIC THEROPODS

By the Jurassic Period, herbivores were starting to reach those enormous sizes that we typically associate with dinosaurs. And as is to be expected, where large prey like herbivores is available, large and skillful predators will inevitably appear to tackle them.

CRESTED PREDATORS

Early in the Jurassic came one of the most unusual theropods, *Dilophosaurus*. This dinosaur shot to unexpected fame when it appeared in the 1993 movie *Jurassic Park*—although, unfortunately, this rise in popularity was accompanied by a host of very persistent misconceptions. *Dilophosaurus* had neither an expanding neck frill, nor the ability to spit venom. It did, however, possess that bizarre double-crest on top of its head. Analysis of the bone shows the crest of *Dilophosaurus* was highly pneumatized, suggesting that its primary function was for display. It was certainly too fragile to have been of any combative use. The crest was also much larger than it is depicted in the movie. Standing up to about 8 ft (2.5 m) in height, *Dilophosaurus* was one of the first truly large theropod dinosaurs, but through the Jurassic it was followed by many more.

Another predator of the Jurassic Period, *Ceratosaurus*, also had distinctive head features—two large bone ridges just in front of the eyes and a large central ridge, which would, in life, have supported a keratinous horn structure.

← The skull of the theropod *Dilophosaurus wetherilli,* showing the impressive vascular crest. The gap between the bones around the nostril indicates that this dinosaur did not have a particularly powerful bite.

ALLOSAURUS

The best known of the Late Jurassic theropods is surely *Allosaurus*, the 30-ft (9-m) long hunter with a narrow snout, those *Ceratosaurus*-style eye ridges, and three particularly vicious-looking claws on the end of each hand. Fossils of this North American apex predator are common in the Morrison Formation (a hotbed of dinosaur discovery across several northern states in the United States), making it one of the most well-studied theropods of all.

Recent analysis of the jaw structure of *Allosaurus* has suggested that it could have opened its jaws to an extreme gaping angle of 92 degrees, and used this in combination with a strong neck action to effectively hack at its prey. This theory remains controversial, though, as the detrimental effects this action may have had on the head of the animal remain unknown.

THE SMALL THEROPODS

While the larger species typically attract all the attention, the true innovators of the Jurassic Period were the small theropods. It was they who made the biggest evolutionary leap by far, which would change the ecosystems of the world forever, but we will discuss these small theropods later in Chapter 7.

GLOBAL DOMINANCE

Thanks to an abundance of fossils, the theropod predators of Jurassic North America may be the best known, but plenty of others have been discovered elsewhere. They include the nimble hunter *Megalosaurus* of Europe, and the aptly named, 23-ft (7-m) long *Afrovenator* of Africa. We even find theropods from Antarctica in the Jurassic. Well before the continent was locked in ice, the single-crested *Cryolophosaurus* hunted here, although due to difficulty accessing fossil sites, so far only the fossils of one individual have been found. Oddly, very few remains of theropods from the Jurassic have been discovered in Australasia, whether due to a lack of fossils found or an absence of theropods we can't be certain (a saying in paleontology being "absence of evidence is not evidence of absence").

THE CHASERS

If ambush fails, hunters may have to run to catch their prey, and some dinosaurs evolved especially for this strategy. Perhaps the most dedicated were the abelisaurids, a group of theropods found across South America, Africa, and India through the Jurassic and Cretaceous.

Abelisaurids had massively reduced arms, to the point where from a distance you probably wouldn't even notice they were there. Their heads were fairly stout compared to the elongate shape of other theropods, and overall their sizes were modest. Most were comparable to an adult human in height, with the exception of some like *Carnotaurus sastrei*, at nearly 10 ft (3 m) tall, which is also known for the distinctive protruding horns of its skull. Abelisaurids probably specialized in chasing small prey, their powerful legs and spring-loaded feet giving them effective bursts of speed—equally useful for scarpering when scavenging the kills of other, larger predators, should they return to reclaim them.

↙ The hugely reduced arms of abelisaurids would have seemed almost entirely useless compared to the arms of other theropods.

↙ The skull of the theropod *Carnotaurus sastrei* shows two large bone protrusions above the eyes. In life, these protrusions would likely have been covered by a keratin sheath.

→ *Carnotaurus sastrei* was one of the largest known abelisaurid dinosaurs and would have been a top predator in Argentina 70 Mya ago. Thanks to exceptional skin preservation, it is known that *Carnotaurus* lacked the feathers seen in many other theropods, and had scaly skin and randomly distributed large "feature scales."

FISHING

When it came to hunting prey, theropods did not restrict themselves purely to what they could find on land. There was no need to stick to that when there were plenty of fish in the waters. Catching fish requires a completely different skillset to that needed for tackling dinosaur prey, and no group evolved to specialize more for this than the spinosaurids. First evolving in the early Cretaceous, this group was hugely successful, with representatives on every continent.

CONVERGING TACTICS

The most distinctive features of the spinosaurids were their skulls, which were very elongate, with long, thin snouts lined with rounded, narrow teeth. This head and tooth shape combination was ideal for snapping up fish and preventing them wriggling free. It's a familiar look, with spinosaurid skulls looking quite like those of crocodiles. The reason for this is called convergent evolution. Essentially, both groups were evolving to solve the same problem and came up with the same solution.

HEAVY CLAW

Converging with our own fishing inventions, some spinosaurids even evolved fish hooks. *Baryonyx* was a dinosaur from the United Kingdom dubbed "Heavy Claw" due to the 12-in (30-cm) sickle claw on each "hand," which was used to snare fish from rivers. A diet of fish is nothing to be sneered at either, for it sustained the largest terrestrial carnivore currently known to science. At 46 ft (14 m) in length, *Spinosaurus*, the dinosaur that gave this group its name, was 6½ ft (2 m) longer than a *Tyrannosaurus rex*.

THE LIFE AQUATIC?

Instantly recognizable from the 5-ft (1.5-m) high sail running along its back, *Spinosaurus* also holds the unfortunate crown of being perhaps the most controversial dinosaur among paleontologists, simply because nobody seems able to agree on how best to reconstruct it. Evidence from the high bone density of its ribs and femur, along with the flattened, paddle-like structure of its tail, have led some scientists to suggest that *Spinosaurus* was an aquatic dinosaur, more comfortable actively swimming in rivers than standing or wading around the banks. It seems that every year more papers are published reassessing the look and lifestyle of *Spinosaurus*, but maybe one day more fossils and study will help reveal the true nature of this enigmatic beast.

← The Brazilian *Irritator challengeri* shows many dinosaurian fishing adaptations—the slender, crocodile-like snout and large clawed arms—athough it lacks the sail of the famous *Spinosaurus*.

THE TYRANTS

There can be no question that the most famous dinosaur of all is *Tyrannosaurus rex*. When this dinosaur was first described in 1910 it was an instant sensation. A combination of early 20th-century American performance and the staggering features of the great predator captured the imagination like no other dinosaur has before or since.

POWERFUL BITERS

Tyrannosaurus rex is just one species from a whole group of dinosaurs, the tyrannosaurids, many of which shared one major trait: an extremely large and powerful set of jaws. Uniquely, the nasal bones of tyrannosaurids were fused, giving their skulls increased resistance to pressure and allowing for a higher bite force. Using evidence of injury in other dinosaurs, and mechanical analysis of the skull, *T. rex* likely had a bite force of over 40,000 Newton, strong enough to break bones. Their teeth were enormous, conical, and had serrations along the reverse edge. They were specifically evolved to take on large dinosaurs.

It has been proposed in the past that *T. rex* was a scavenger, not a hunter. However, we do have evidence of healed wounds made by *T. rex* in other dinosaurs, suggesting that they were attacked while still alive. It seems likely therefore that the animal was an active hunter.

← The 5-ft (1.5-m) long skull of *Tyrannosaurus rex* evolved specifically to have a strong bite force. The extreme stresses involved in biting were dissipated by the shape of the skull.

TINY ARMS

It is the impressive skull that was the reason for the tyrannosaurids' other famous feature, their tiny arms. If they'd had arms the size seen in other theropods, they would have been too front-heavy to balance effectively, and so evolved smaller arms in order to maximize their head strength. Whether these arms had any function whatsoever is still a matter of some scientific debate.

FEATHERED REX?

Whether *Tyrannosaurus rex* had feathers is a hard question to answer conclusively. However, fossils have been found with preserved *Tyrannosaurus* skin, and from this we can deduce that at least some sections, including the legs and head, were covered in scales and not feathers. While evidence for *T. rex* leans toward a mostly scaled appearance, we know some tyrannosaurids did have feathers. An early member of the group, *Yutyrannus*, from northeast China, shows direct evidence of a feathered body from fossilized imprints, which makes this 30-ft (9-m) long theropod the largest animal with confirmed feathers yet described by science.

WORLDWIDE FAMILY

Tyrannosaur ancestry can be traced back through to the Jurassic, with species found across much of the Northern Hemisphere. The earliest group members, such as *Guanlong* of Late Jurassic Asia, look remarkably different to later members, even sporting a large head crest. But it is in the Late Cretaceous of western North America that the group truly flourished. The tyrannosaurids *Daspletosaurus* and *Gorgosaurus* are the top predators found in the fossil-rich Dinosaur Provincial Park of Canada. Another Canadian member, *Albertosaurus*, has the honor of being named the National Dinosaur of the country. And to the south, in what is now the United States, *T. rex* was king.

GIANT KILLERS

T he tyrannosaurids were not the only giant dinosaur hunters. The Cretaceous was also a golden age for other terrifyingly large terrestrial predators. In South America, for example, even the sauropods were not safe from the 43-ft (13-m) long *Giganotosaurus*, while in northern Africa, *Carcharodontosaurus* ruled, taking their name from their thin, shark-like teeth. Neither had jaws quite as powerful as those of *Tyrannosaurus rex*, but both were larger, and they had more formidable clawed arms that could be used against prey.

Both *Giganotosaurus* and *Carcharodontosaurus* were part of the Allosauroidea family, which had dominated the Jurassic. We know this because of the ridge structure on each side of their heads, running from the nostrils to the eyes. This bumpy bone structure, known as rugose texture, suggests the presence of keratin on the surface, extending the crests to more exaggerated display structures. They possibly used these to attract mates, providing a glimpse into a more intimate side of the lives of these huge predators.

↓ *Giganotosaurus carolinii* had a skull measuring over 5 ft (1.5 m) in length. It is the largest known theropod from South America. The top surface of the skull shows the bumpy rugose texture, which helps distinguish it from that of other large theropods such as the tyrannosaurids.

→ *Carcharodontosaurus saharicus* was one of the largest terrestrial predators ever to have existed. Although its skull and teeth are relatively well known, there are not many known remains of the rest of its body, with much material having been lost in the Second World War.

HUMBLE BEGINNINGS

T he sauropod dinosaurs (often referred to as the "long necks") were the largest dinosaurs of all, the giants of the ancient world. However, tracking back through their family tree, it can be seen that even these huge creatures started off small.

BASAL DEBATE

Sauropodomorphs are saurischian dinosaurs, meaning that despite drastic differences in appearance, they are more closely related to the theropods than any other major herbivorous group. The similarities are more evident in the group's earliest members.

One particularly controversial potential relative is *Eoraptor*, 230 Mya, a small bipedal omnivore among the earliest dinosaurs. The exact placement of this dinosaur is debated—was it a basal theropod or sauropodomorph? This confusion shows just how similar the first sauropod relatives were to early theropods.

LATER TRIASSIC

By the end of the Triassic, the early sauropodomorphs were clearly on their own evolutionary path. Studies of the teeth of *Thecodontosaurus*, an early sauropodomorph from the United Kingdom, show a transition from carnivore to herbivore, although it was still a cursorial biped.

Elsewhere in Europe the sauropodomorphs started to become larger, as evidenced by the nearly 33-ft (10-m) long *Plateosaurus*. The neck of *Plateosaurus* was longer than in most contemporary dinosaurs of the time, although still short by sauropod standards. And, as with all the Triassic sauropods, it was a biped.

FORELIMB USE

With their forelimbs free, several of the early sauropodomorphs had evolved large claws for feeding. It's been suggested that these arms would have been used to rip foliage from trees but may also have served in defense.

TREE CONFUSION

As evidenced by *Eoraptor*, the base of the sauropodomorph family tree is unclear. The problem reaches beyond this and has resulted in many examples of "wastebasket taxa." This is when many different species are accidentally lumped into one. *Plateosaurus* is one of the largest offenders here, with no agreement on which of the many assigned specimens are legitimately in the same genus. Surprisingly, the reverse can also be true, when two species that should be the same are classed as different. This has been argued for *Thecodontosaurus* and the Welsh *Pantydraco*. It stems from a lack of consensus on what the true defining characteristics of each should be.

← *Plateosaurus* is known from many specimens across Europe, making it probably the most extensively studied of all Triassic sauropods.

ON ALL FOURS

As sauropodomorphs evolved to reach larger sizes, there came a point when they could no longer balance effectively on just two legs and needed to transfer to a fully quadrupedal posture. There is, however, some debate over when exactly this transition took place. The traditional view held that the first quadrupedal sauropodomorphs evolved in the middle of the Jurassic. But more recent fossil evidence suggests an earlier evolution.

Ledumahadi, the largest known animal from South Africa, is thought to have weighed 13 US tons (11,800 kg) and walked as a full quadruped during the earliest stages of the Jurassic. Footprints from Greenland hint at an even earlier origin of quadrupedality, into the Late Triassic, although no skeletal evidence is confirmed as associated with these tracks. Some of the sauropodomorphs that came after *Ledumahadi* show bipedal stances, suggesting that the evolution of quadrupedality may have occurred multiple times in the sauropod family tree. Essentially, the group was exploring different ways of achieving massive body sizes.

↓ This trackway in Greenland appears to show a dinosaur moving quadrupedally in the Late Triassic, but the maker is a mystery.

→ Although smaller than its later relatives, it is thought that when *Ledumahadi* existed, 200 Mya, it may have been the largest animal ever to have lived.

Shortly after, other African sauropods, like *Vulcanodon*, are seen to have evolved quadrupedality with fully columnar legs.

SAUROPOD MORPHOLOGY

B y the Late Jurassic, the sauropods had evolved to become the largest animals ever to have walked the Earth. How they accomplished this is a marvel of biological engineering, but they largely achieved it by making themselves as light as possible.

LIGHT BODIES, LONG NECKS

Some sauropod dinosaurs weighed more than 50 US tons (45,000 kg), but this is actually lighter than they should have been. This was made possible due to huge air sacs throughout their bodies. It's estimated that internally some of these giants may have been up to 10 percent air. Having these pneumatic structures made it easier for the sauropods to move because their muscles did not have to expend as much energy for simple movements.

Classically, sauropods were reconstructed with a "swan-neck" pose, a high, curving "S" shape thought to have allowed them to reach the tops of trees. Reaching leaves that no other herbivore could was certainly part of the neck's function, but the positioning is more debated. Going for the greatest verticality required necks to be heavily muscular, as is seen in *Giraffatitan* and *Brachiosaurus*. These dinosaurs both had much larger forelimbs in comparison to their hind limbs, allowing an even greater reach.

↙ *Supersaurus* shows the classic sauropod body plan: four columnar legs with a long neck balanced by a long tail.

SKULLS AND FEEDING

Sauropod skulls were very small compared to the rest of their bodies, which is likely related to how they fed. Their peg-like teeth were not adapted for chewing food, so they did not need heavy jaw musculature.

Diplodocus-like sauropods have narrow skulls, which are relatively flattened compared to the stout skulls of the macronarian sauropods, like *Brachiosaurus*. The rounded high-tops of macronarian skulls were formed by arching nasal bones, creating a huge skull cavity (although the nostrils themselves remained at the front of the snout, as expected). This large cavity may have helped the animals communicate through sound, or been part of a modified respiratory system suitable for their size. The answer isn't definitively known.

Tooth analysis suggests that different sauropods specialized in feeding on different foliage. This niche partitioning allowed multiple species of giant herbivores to survive together in one environment.

TAIL FUNCTION

Some *Diplodocus* have been reported as having 80 caudal vertebrae, giving them exceptionally long tails. Possible functions of this tail have been suggested, beyond just their use in balancing the animal. Early theories suggested the tail may have been used as a weapon, although its very thin end couldn't have transmitted much impactful force. An exception to this may have been *Shunosaurus*, whose tail ended in a club-like collection of spiked osteoderms (essentially armored scales within and on the skin). Another theory suggests that the tails were used for visual display and may have had patterning along them.

RECONSTRUCTION MYTHS

All scientists agree that the old notion that sauropods were in some way aquatic, needing water to support their body weight and using their necks as snorkels, is false. They could support their weight on land and the pressure differences and difficulties in staying submerged would have made life underwater impossible. Another extreme theory asked whether sauropod tails could be flicked to generate a "whipcrack" sound for signaling. But any modeling of the negative effects this would have had on the soft tissue around the tail would refute that as a plausible function.

THE TITANS

During the Cretaceous, the largest dinosaurs were the perfectly named titanosaurs. Included in this group and time were the biggest terrestrial animals the planet has ever seen. Due to the patchy fossil record for certain rarely preserved parts of their bodies, paleontologists have had to rely on specimens of their embryos, still preserved within the eggs, to fill in the blanks.

ARMORED SKIN

Titanosaurs are the only sauropod group seen to have osteoderms—nodules of bone set within the skin of the dinosaur, usually located across the back. In the small (relatively) titanosaur *Saltasaurus*, some of these osteoderms could reach 4³/₄ in (12 cm) across. It has been assumed that these were used for defense, raising the possibility that even such leviathans were not safe from large theropods, although with so little evidence to go on, this is still uncertain and it's possible they served another unknown function.

TOE LOSS

A more subtle difference can be seen in the front limbs, with later titanosaurs appearing to lack any kind of claws. It is thought that these were lost as vestigial organs (they no longer served any function). This loss was not universal across titanosaurs, however, with some species maintaining front toe claws. It's likely that there wasn't a particularly high selective pressure on this feature, and so evolution acted slowly. Trackways from Australia show the presence of front claws, whereas their counterparts in South America are lacking, hinting at different rates of evolution of this feature across the globe. All titanosaurs do, however, maintain the claws on their back feet.

WHO'S THE BIGGEST?

Whenever dinosaurs are discussed, it is inevitable that the question should be asked: Which species of dinosaur was the biggest of all? It's a surprisingly tricky question because, although it can be said with some certainty that it was a titanosaur of some kind, if you try to narrow it down any further, things get murky. Estimations of body weight are precisely that, estimations. There's plenty of room for error, as well as the human temptation to exaggerate finds to maximize media coverage. The current best estimates would place *Argentinosaurus* on top of the podium, weighing in at an impressive 83 US tons (75,300 kg). However, it's always sensible to take such statements with a pinch of salt.

↘ Finds of huge collections of titanosaurs and their nests in South America indicate that these huge sauropods likely traveled in herds.

POOR PRESERVATION

Despite some sauropods having individual bones larger than an adult human, they have a relatively bad fossil record compared to other dinosaur groups, and their size is part of the problem. This is because one of the factors determining good fossil preservation is the speed of burial of the remains. For small organisms this is relatively easy, but a huge amount of sediment displacement is required for sauropods. Instead of being quickly covered, sauropod remains were regularly left on the surface for scavengers to break apart. The air sacs that kept sauropod bones light also made them fragile when exposed and liable to be crushed and destroyed by the pressures of the fossilization process

Finally, a surprising reason for their rarity is due to us as extractors. Small fossils are easily removed for study, whereas giant sauropod bones can require heavy machinery and days of excavation for just a single specimen. In some ways, there is an unfortunate bias against their study in paleontology.

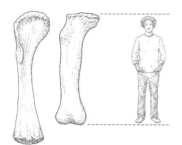

← Sauropods had some of the largest individual bones of any animal. Single limb bones, such as the femur and humerus, could be longer than the height of an adult human.

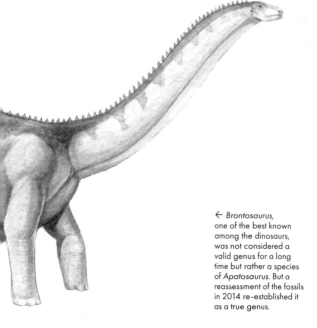

← *Brontosaurus*, one of the best known among the dinosaurs, was not considered a valid genus for a long time but rather a species of *Apatosaurus*. But a reassessment of the fossils in 2014 re-established it as a true genus.

OTHER GIANTS

T he sauropods were the largest of all dinosaurs, but they were by no means the only group to achieve such huge sizes. Gigantism is seen to have evolved repeatedly in the dinosaurs, across many of the disparate groups.

ORNITHOPODS

Other herbivores became large enough that they were comparable to the sauropods. For example, weighing an approximate 14 US tons (12,700 kg) and with a body length of 50 ft (15 m), the ornithopod *Shantungosaurus* was the largest known dinosaur outside the Sauropoda. Part of the hadrosaur family, this dinosaur was similar to the better-known *Edmontosaurus*.

Shantungosaurus was not alone among the hadrosaurs either, as other genera, like *Saurolophus* and *Magnapaulia*, reached similarly enormous sizes. Hadrosaurs have a relatively good record of fossil skin and so we know they had a scaly appearance. Feathers first evolved to provide insulation in dinosaurs, but it would seem the larger bodied dinosaurs had less use for them and may have secondarily lost them.

It is thought that hadrosaurs were able to achieve such large sizes through continuous growth, as opposed to the bursts of seasonal growth seen in most other dinosaurs. They likely lived quite long lives by dinosaur standards, much the same as large mammals are seen to have relatively long life spans today.

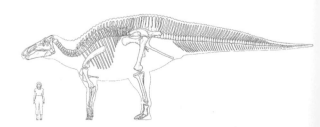

OTHER HERBIVORES

The largest of the horned dinosaurs may not have been *Triceratops*, but a close relative, as evidenced by the similar name: *Eotriceratops*. The skull alone of this dinosaur was almost 10 ft (3 m) in length, from the tip of the beak to the back of the frill. Likewise, *Stegosaurus* could potentially have reached 26 ft (8 m). It is unlikely that these herbivores attained the length or height of a double-decker bus (as is often suggested), but this does not detract from just how massive they were.

CARNIVORES

As we have seen, the largest known carnivorous dinosaur was the 46-ft (14-m) long *Spinosaurus*. Just among the theropods, examples of gigantism are seen to have evolved independently multiple times. This includes the tyrannosaurids and several other members of the Coelurosauria. What is especially important, however, is their stance, as these huge theropods were the largest ever known fully bipedal animals. For the most part, evolution to large sizes was likely a response to prey animals doing the same. Predators and prey are invariably linked to evolve together, and size is just one factor in the evolutionary arms race.

MESOZOIC GIGANTISM

It wasn't just the dinosaurs that achieved massive sizes during the Mesozoic Era. Other reptile groups did too. Notable among these were the ichthyosaurs, the largest of which was *Shonisaurus*—estimated to have reached up to 50 ft (15 m) in length. That's about the same as a modern Sperm Whale (*Physeter macrocephalus*), and the presence of very small teeth in its long jaws suggests it may have fulfilled a similar niche in its environment. Although the ichthyosaurs have since been "overtaken" by whales as the largest creatures in the oceans, the dinosaurs still hold the record for the largest terrestrial animals ever and azhdarchid pterosaurs the largest animals to have flown.

← At about the size of a double-decker bus, *Shantungosaurus* showed that sauropods weren't the only giant herbivorous dinosaurs.

SAFETY IN NUMBERS

Perhaps the simplest form of defense is to go by the adage of safety in numbers. The principle is simple: If you are surrounded by many other animals, the odds of you being the one singled out by a predator are substantially reduced. As such, we see herding behavior in animals the world over today. But was the same true of dinosaurs? In short, yes. Evidence for herds of dinosaurs can be found throughout their reign. The most direct evidence comes from trackways, such as in Denali National Park, in Alaska. Here we find many matching sets of footprints, clearly from the same dinosaur species (in this case, hadrosaurs) and all moving in the same direction.

Mass assemblages of dinosaurs caught in storm surges, like the centrosaurs of the Hilda Mega-Bonebed, in Canada, suggest they may have been migrating together. From a distance, some Mesozoic scenes may have looked much like those seen on the open grasslands of today.

↓ The classic three-toed shape makes theropod footprints instantly recognizable. They strongly resemble the imprints of modern birds.

↓ Large sauropod footprints are rounder, the toe imprints showing the direction of travel. These footprints indicate that the sauropods were traveling in herds, nearly all in the same direction.

→Lacking any specialized weaponry, dinosaurs such as hadrosaurs would have relied on herding behavior for defense. Fossilized trackways and the remains of nests from this dinosaur group have been found across the world.

HORNED DINOSAURS

A popularly reconstructed dinosaur battle is that between *Tyrannosaurus rex* and *Triceratops*. The three-horned skull of *Triceratops* lends itself to this idea on initial interpretation, with the frill depicted as protecting the vulnerable neck of the animal. However, evidence from across the family to which *Triceratops* belonged, the ceratopsians, suggests a more complex functionality.

EARLY CERATOPSIANS

Originating in the latest Jurassic, the first distinctively ceratopsian features can be seen in the early Cretaceous *Psittacosaurus*. Small and bipedal (unlike later large quadrupeds), the head shows a horned beak and other keratinous structures below the eyes. These are possibly related to the frill ornaments and horns of later ceratopsians.

Thanks to excellently preserved specimens from the Jehol Province of China, paleontologists have been able to reconstruct this animal in incredible detail. Long, quill-like feathers ran along the tail and the skin was camouflaged with black patterning. Even the cloaca of the animal can be seen. This single genital opening was long presumed to be in dinosaurs, but the *Psittacosaurus* fossil was the first time its presence had been confirmed.

FACIAL WEAPONRY

Many later ceratopsians sported a diverse array of forward-facing horns. Some had the two familiar large, diverging prongs (like *Triceratops*). Others had a singular rhino-esque spike (as in the case of *Styracosaurus*). And a few had flat surfaces of dense bone above their nostrils (namely, *Pachyrhinosaurus*).

The obvious thought is that they used these appendages in defense against predators. Wounds inflicted by theropods are common in ceratopsian specimens, although the fact that many show signs of healing implies they had some way of fighting back to escape death.

However, the primary use of these horns would have been in competition against each other. The spacing of the horns on a *Triceratops* skull seems to lock together perfectly, in a way that would support their use in head-on-head combat, much like modern deer clashing antlers during the rutting season. The glancing blows and scars often seen on ceratopsian skulls lend further support to this idea, as well as evidencing what a bloody affair it could be.

FRILL FUNCTION

The large frills of ceratopsians would almost certainly have been used in a form of display, be it as a show of strength to an attacking predator or as a proof of fitness to potential mates. Several species show ornamentation around the edges of their frills (parietal spikes). *Triceratops* sported an array of small uniform structures, whereas others, like *Styracosaurus*, had much more elaborate spines, the largest of which could be about 2 ft (60 cm) in length. Despite looking rather fearsome, their awkward positioning means they likely had no function beyond display.

SPECIES ONTOGENY

It was once suggested that multiple different species of ceratopsian were actually just the same species at different stages of life. The idea was that as *Triceratops* aged, the bone density of the frill would lessen, eventually opening up massive holes, as seen in a contemporary species of the time, *Torosaurus*. While changes to morphology through aging (known as ontogenetic variation) is seen in animals and poses potential issues for dinosaur paleontology, this particular theory is unlikely. The lack of evidence and other features separating the two groups clearly shows them as distinct.

→ The impressive skull of *Triceratops* can measure nearly 10 ft (3 m) from the beak tip to the back of the frill.

ARMORED DINOSAURS

O ne feature that has evolved repeatedly in terrestrial vertebrates is body armor. The principle is simple: If your skin is tough, predators will not be able to bite through it. We see this in animals like armadillos, but their best efforts pale in comparison to the ankylosaurs, possibly the most elaborately armored vertebrates of all time.

EARLY REPRESENTATIVES

Early relatives of the ankylosaurs (such as *Scelidosaurus*) were likely facultative quadrupeds, primarily on four legs but able to switch to two in short bursts. Their armor was ornate with spikes, a further deterrent to predators. This is a group feature seen from the very beginning, with the oldest known ankylosaur, *Spicomellus*, of Jurassic Morocco, covered in them. These spikes were also seen in the nodosaurs, a specialized group of ankylosaurs. The American *Sauropelta* had increasingly large spikes along its neck, maxing out with a 2-ft (60-cm) long pair at the shoulders. As is often the case with such extreme features, it's thought these likely evolved through sexual selection.

PRESERVATION POTENTIAL

One of the best preserved large dinosaurs ever to be discovered is the 18-ft (5.5-m) specimen of *Borealopelta* found in Alberta, Canada, in 2011. The head and body are preserved in a lifelike position, including even the keratinous covering of armor, just above the underlying bone. From the right angle, it looks to be merely sleeping. Rather morbidly, the reason for this exceptional preservation is thought to be the weight of the armor potentially drowning the animal after it was swept into the sea. Being so top-heavy meant the dorsal side was quickly buried in the sediment before it could be disturbed by scavenging sea life.

TAIL CLUBS

Being an ankylosaur was not all about passive defense, as later species evolved impressive weaponry. *Ankylosaurus*, the largest of the group, had a solid club on its tail, possibly for use against predators like *Tyrannosaurus rex*, whose bites could test even the toughest armor. This club was made of two extremely large osteoderms. Surrounding it were several smaller bone structures, covering the final few vertebrae of the tail. It's estimated that, by swinging this tail, the animal could have generated a force of 4,800 Newtons per second—essentially the same as being hit by a mid-range car.

Although it is satisfying to imagine the tail club of ankylosaurs being used in battle against a theropod enemy, there is little evidence to confirm its function as a weapon at all. If it were being used in this way, you would expect to see evidence in the bones, although initially none were seen. It wasn't until 2021 that a specimen of *Tarchia*, an Asian ankylosaur, was reported to have asymmetrical bone growth in the club due to repeated striking behavior. But this still doesn't confirm whether the club's primary use was against attackers, competition with other *Tarchia*, or as a means of display against inanimate objects, perhaps to create sound. Yet this may be a case of limited data—there's only one known fossil tail club of *Ankylosaurus* and a handful of the similar relative *Euoplocephalus* to go on.

↙ The tail club of *Ankylosaurus* was almost 20 in (50 cm) wide. Made from bone, tail clubs could have transmitted huge forces.

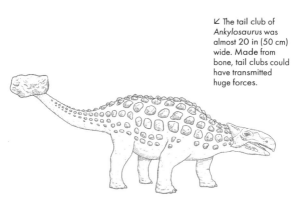

THE STEGOSAURS

T he iconic body shapes of the stegosaurs make them instantly recognizable. These quadrupeds had backs lined with geometric plates of bone, small skulls (their brains only about the size of a walnut), and a tail ending in formidable spikes. In the largest species of *Stegosaurus*, these spikes could reach over 3 ft (1 m) in length.

TAIL SPIKE DEFENSE

Unlike the tails of ankylosaurs, we do have proof that these tail spikes were used to fend off predators. One specimen of *Allosaurus* shows a tail vertebra with damage that perfectly matches the shape and size expected were it to have been punctured by a *Stegosaurus* tail.

Not all stegosaurs had spikes only on their tails. The spikes of the north African *Kentrosaurus* ran the entire length of the tail and beyond the hips. In addition, they also had two lateral spines extending from their clavicles. This dinosaur clearly felt the need to be protected from all angles.

PLATE FUNCTION

A long-standing debate in paleontology surrounds the function of stegosaur plates. Lined with blood vessels, it has been suggested these plates could have been used for thermal regulation (pumping them with blood to act as solar panels or to cool off), or as a form of display when flushed with color. A common misconception for *Stegosaurus* is that the plates were paired in two symmetrical rows. In fact, the parallel rows were off-set, meaning both plate rows would be visible regardless of which side the animal was viewed from. This has been suggested as evidence for the display theory, with the plates forming a larger uniform surface.

DIMORPHIC DINOSAURS?

Adding to the controversy surrounding the function of stegosaur plates, it has also been suggested that stegosaurs showed evidence of sexual dimorphism (physical differences between the sexes), which is a topic for debate among paleontologists (see Chapter 10, page 124). One study examining the femur robustness of multiple *Kentrosaurus* specimens seemed to suggest two clear morphs: one robust and one more gracile. One potential reason for this could be individuals of different sexes having different builds, although with only bones to go on, it's hard to prove this for certain. This has also been suggested for the plates of *Stegosaurus*, with two potential morphs possibly indicating a sexual dimorphism: some animals showing rounded plates and others being tall and angular.

→ *Stegosaurus* had a back lined with ornate bone plates, the largest of which were directly over the hip and could reach around 2 ft (60 cm) in height.

THICKENED SKULLS

Members of the dinosaur group Marginocephalia were characterized by specialized bone structures around the edge of their skulls. They include the ceratopsians with their frills and also the pachycephalosaurs, which had incredibly thick domes of bone on top of their heads, bordered by small spike structures.

In some species, such as *Pachycephalosaurus wyomingensis*, the bone could be over 9½ in (24 cm) thick and capable of transmitting large forces on direct impact. This would be perfect for dissuading predators from a meal, although like the ceratopsians, it is believed their primary use was for competition with each other.

Much like modern Bighorn Sheep (*Ovis canadensis*), the impacts created by these 13-ft (4-m) long animals would have been huge and, even with the effective impact absorption of the bone, may have caused traumatic injury to the animal's brain. To mitigate this, "flank butting" rather than head-on-head combat has been suggested, although injuries to the skull suggest they also took part in the latter.

↓ Ornamented with bone spikes and nodules, the spectacular skull of the herbivore *Dracorex* has often been likened to that of a mythological dragon.

↓ *Pachycephalosaurus* had an extremely dome-shaped skull, giving the animal a very distinctive look. The head also features several other bone structures.

→ Pachycephalosaurs were generally quite small dinosaurs, the largest being only 5 ft (1.5 m) tall. However, their thick skulls meant they were more than capable of defending themselves. As such, the group thrived in the Northern Hemisphere during the Late Cretaceous Period.

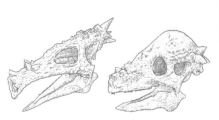

SPEED

In a fight-or-flight situation, taking the latter option always offers the best odds for animals. This is because fleeing is less energetically expensive and carries a lower risk of injury. For the larger and heavier dinosaurs, this was not a possibility, but for others it was a specialty.

CALCULATING DINOSAUR SPEED

Neither birds nor crocodiles move exactly how dinosaurs did, meaning inventive methods of calculating speed in dinosaurs is required. Evidence can come from biomechanical analysis of bones, muscle reconstructions, and estimates from dinosaur trackways. Attachment "scars" on the bones help give an idea of the size of muscle and allow for relative comparisons with similar living organisms. If the size of the animal making the footprint is known, then the stride length can allow for an estimate of the speed at which it was traveling. The shape of the footprint and force of impression left by the toes can also be used to inform whether the animal was running or walking.

~ Biomechanics ~

Using Finite Element Analysis (an engineering technique used to test vehicles and architecture), the load-bearing properties of the bones can also be studied. Determining the dissipation of stresses and the ability to withstand certain forces allows for the setting of hard maximums for the animal's possible speed. The same technique can also be used to calculate things like bite force and jumping potential. Using these techniques, it's thought that the fastest dinosaurs (ornithomimids) could probably reach speeds of up to 28 mph (45 km/h), while the large predatory theropods likely capped out at 20 mph (32 km/h).

ORNITHOMIMIDS

The name ornithomimid translates to "bird-mimic," which is precisely how their skeletons appear. Narrow skulls with beaks are perched atop long, slender necks and lightweight bodies with long tails and powerful legs. Fossil evidence from *Ornithomimus* shows just how like an ostrich they were, having a coat of feathers transitioning to bare-skin legs from the thigh down which allowed them to run unhindered. The two groups were also comparable in height, although the largest ornithomimid, *Gallimimus*, could reach up to about 20 ft (6 m) in length. Despite a lack of teeth, ornithomimids were theropods, although they stuck to a more generalist diet, likely feeding mostly on plants.

DINO-BIRD LINK

L ess than 20 years after first being named by science, the connection between dinosaurs and birds was already being noticed. As new species of dinosaurs were discovered, the obvious differences of toothed jaws and bony tails were fast being overshadowed by their similarities, hinting at a shared family tree.

Not only do dinosaurs and birds share certain unique bones (such as the furcula, which is commonly known as the wishbone), but the internal structure of the bones is also similar. Both have "hollow" bones, a weight-saving trait that allowed dinosaurs to grow to large sizes and allows birds to maintain flight.

After a lapse in popularity of the dino-bird theory, an explosive find in 1996 brought this back to the forefront of research when *Sinosauropteryx* was discovered. This small theropod was the first dinosaur fossil to show clear evidence of feathers. The initial trickle of evidence became a flood as more and more feathered dinosaurs were discovered and the link to birds is now in no doubt.

↙ *Velociraptor* was a lot smaller and more bird-like than is usually portrayed in the media. Quill knobs found on the forearm of the famous predator confirm that it had feathers. They were likely for display, as it lacked the musculature and large wingspan to enable it to fly.

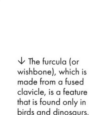

↓ The furcula (or wishbone), which is made from a fused clavicle, is a feature that is found only in birds and dinosaurs.

↓ Dinosaurs weren't restricted to just downy feathers, as fossils from the Mesozoic show the same structures as those seen in feathers today.

RAPTORS

Alongside *Tyrannosaurus rex*, the raptors are probably the most famous of all predatory dinosaurs. The formal name for this group is the Dromaeosauridae, part of a larger group of theropods known collectively as the paravians. As the name suggests, these dinosaurs are extremely closely related to birds, sharing many features.

SICKLE CLAWS

The most infamous feature of the raptor dinosaurs are the sickle claws on their inner toes. The enlarged size of the claw compared to the others, and its extreme curvature (much like modern eagle talons), has led to speculation that it would have been used in hunting. One remarkable fossil found in Mongolia in 1971 lends support to this theory. The fossil find, known as the "Fighting Dinosaurs," shows a *Velociraptor* locked in the act of fighting a *Protoceratops*. The claw of the raptor appears to be embedded in the neck of the ceratopsian, suggesting it may have been used for targeting vulnerable areas to secure a kill. The *Protoceratops* seems to have caught the arm of the raptor in its jaws and trapped it underneath its body weight, before both were buried suddenly by collapsing sand. Other theories suggest that raptors leaped onto their prey and used their claws to grip on— although this means the claws would need to have been able to withstand high levels of strain, possibly beyond their limits.

← The sickle claw of raptors is their most famous feature. In life it would have been covered by a keratin sheath, making it look even more formidable.

FEATHERED HUNTERS

Contrary to how raptors are frequently depicted in the media, they would have been covered in feathers rather than scaly. Direct evidence for this has been found on the forelimbs of *Velociraptor*, which show quill knobs—attachment sites for large pennaceous feathers (the same kind of fully developed feather seen in the outer layer of modern birds). It's believed that feathers first evolved in dinosaurs for insulation (essentially being highly modified keratin structures able to "fluff up" and regulate body temperature more effectively). The large feathers seen in raptors, however, likely also had other functions. They may have been brightly colored for display or even served a use in hunting.

SIZE DISPARITY

Another regular misconception surrounding the raptors concerns their size. Although some, such as *Utahraptor*, could nearly reach the height of an adult human, most were much smaller. The famed *Velociraptor* was in reality only about 2 ft (60 cm) tall. However, even with their small size and feathers, raptors were far from being capable of flight. Their feathers probably acted as stabilizers if the animal were to jump on prey. The stiff rod extensions on their tail vertebrae would also have acted as stabilizers, allowing the animal to balance effectively in such high-speed maneuvers.

PACK HUNTING

Isotopic evidence from *Deinonychus* teeth suggests that raptors shifted their diets as they grew, transitioning from very small prey items to other dinosaurs in adulthood. This idea goes against the somewhat popular idea of raptors as sophisticated pack hunters. If in a complex social group, the young would likely have fed on the prey caught by adults, not their own unique diet.

The idea of raptor pack hunting gained popularity following the discovery of several *Deinonychus* specimens alongside the larger herbivore *Tenontosaurus*. But this in isolation does not prove anything about how the animal may have hunted. Such behavior is notoriously hard to show from fossils alone.

FOUR-WINGED FLIERS

B irds were not the only dinosaurs to evolve flight. Other theropods also evolved feathered wings that were capable of supporting their body weight in the air, although some did so with a four-winged body plan, their hind limbs also showing long, feathered wings.

SMALLER SIZES

Some of these flying dinosaurs were among the smallest of them all. The best known, *Microraptor*, had a wingspan of 3 ft (90 cm), which is about the same as a modern-day Peregrine Falcon (*Falco peregrinus*). In contrast, the largest of the flying dinosaurs (not including birds) was *Changyuraptor*, which measured a little over 4 ft (120 cm) from head to tail, making it comparable with a goose. This is perhaps still small by other dinosaur standards but large for a flying bird.

POWERED FLIGHT

There is debate as to whether *Microraptor* and its relatives were capable of powered flight, or were purely gliders. Current theories suggest that most four-winged fliers lacked the musculature to flap in the same way as birds, although *Microraptor* itself potentially could. If correct, this might mean that powered flight evolved three or more times within the theropods. It should be noted that only the forelimbs would have been used for powered flight. There was no way for *Microraptor* to "flap" its winglike legs. Instead, these wings would have been used to provide extra lift and also to act almost like ailerons, directing the flight. The fan of feathers on the tail likely served a similar function.

→ With jaws filled with small, sharp teeth, *Microraptor* likely had a diet of insects and small vertebrates. It would have used its wings to travel through the Cretaceous forests.

JEHOL BIOTA

The epicenter of discovery for dinosaur flight has been the Jehol Province of China. During the Early Cretaceous, the landscape of this region was dominated by wetlands. Dense forests with marshy ground that is difficult to traverse, this environment lends itself to the evolution of flight. Small theropods evolved to glide between the trees, finding what they needed in the canopy. Not only are the dinosaurs here unique, but they are also preserved in excellent detail. Many of the specimens are articulated, the full extent of their feathered wings on display. This is possible thanks in part to fast burial by the volcanic ashfalls that periodically swept over the area.

MESOZOIC BIRDS

Although the idea that birds originated from dinosaurs is well known, they did not emerge as a result of the others' extinction. Birds coexisted with dinosaurs for millions of years, although they were in many ways different from those we see today.

ENANTIORNITHINES

The most common and diverse birds throughout the Mesozoic were the enantiornithines. These animals might appear to be regular birds, with full wings and reduced tails, but they retained certain dinosaur characters, such as toothed jaws rather than beaks and also claws on their forearms.

So far over 80 species of enantiornithines have been described, although because bird bones are so fragile and prone to destruction, some of these are based on extremely fragmentary evidence. It has been argued that they were not such efficient fliers as modern bird lineages, although they were still able to ground-launch and perch in much the same way. This ability was enough for them to be the dominant bird group for about 70 million years during the Cretaceous.

ARBOREAL ORIGINS

For many years, the origins of avian flight were debated, with two main theories: "ground-up" and "trees-down." The ground-up origin relied on the dinosaurs developing powerful jumping abilities, with flapping a potential adaptation of using arms to grasp prey or, alternatively, to run away from predators.

Many of the paravians show extensive adaptations to a life in the trees. Their claws show high re-curvature, which would have made them effective for climbing. Their eyes are large and positioned to best judge depth-perception for their landing targets. It's easy to imagine how jumping between trees could take them down an evolutionary path to gliding and then full flight.

ARCHAEOPTERYX

In 1861, one of the most significant fossil discoveries of all time was made in a limestone quarry in Solnhofen, Germany. A single feather, followed by a near-full skeleton, seemed to confirm the presence of birds in the Jurassic Period, 150 Mya. The specimen, which showed both bird and reptilian features, was named *Archaeopteryx*, meaning "Ancient Wing."

The exact position of *Archaeopteryx* in the family tree of birds and dinosaurs is debated, and it is doubtful whether the initial feather discovered is even related to the *Archaeopteryx* skeleton, but the impact the fossil had on our understanding of bird-reptile relationships (and Darwin's theory of natural selection) cannot be understated.

PTEROSAUR COMPETITION

For half the Mesozoic Era, the skies were shared by the pterosaurs and birds. This raises the question of whether they interacted and competed for resources, driving evolution in each other. One theory claims it may have been the arrival of birds that sparked the evolution of gigantism in pterosaurs, as they moved to claim different niches unavailable to the small-bodied early birds.

Avoiding competition would certainly have benefited both groups, and analysis of the functional traits of the jaws of pterosaurs and birds suggests they did just that. So the skies were truly big enough for both groups to survive during the Cretaceous, and had the pterosaurs not become victims of the Cretaceous-Paleogene (K-Pg) extinction event—a sudden mass extinction around 66 Mya of three-quarters of the Earth's animal and plant species (see Chapter 9, page 110)—they may well have continued to share the ecosystem to this day.

→ Found in 1874 by a German farmer, the "Berlin Specimen" of *Archaeopteryx* is the most complete specimen, and was sold by its finder to buy a cow.

WATER BIRDS

Not all early birds were so keen on staying in the air. Early in their evolution came a split in the family tree between those that would become "classic" birds and those that chose a life in the water. This second group was known as the hesperornithines, and they are truly unique creatures.

Without the extreme pressures of reducing weight that flight brings, the hesperornithines retained their bony jaws and teeth. Specifically, they had thin, needle-like teeth that were ideal for snapping up wriggling fish from the water. Today diving birds generally use their wings to propel themselves (penguins are a good example). However, the hesperornithines relied entirely on their legs, which were positioned extremely far back on their bodies and also muscular in order to provide plenty of thrust. Larger species lost their wings altogether, maximizing streamlining.

The hesperornithines were certainly successful during the Late Cretaceous, but unfortunately they could not escape the same fate as the dinosaurs, going extinct some 66 Mya.

↓ The striking skull of *Hesperornis* reveals a clear mix of bird and reptile features. At first glance, the skull resembles that of

modern water birds such as grebes, but the teeth and jaw structure indicate that this animal could have handled a different diet.

→ Hesperornithines were the first birds to adapt to an aquatic lifestyle, and were widespread across the Northern Hemisphere. Most lived in the seas, but some would later evolve to live in fresh water, too.

SURVIVORS

T he word dinosaur can be misleading. When heard, it is usually in reference to the giant reptiles of the Mesozoic, which went extinct 66 Mya. But the dinosaurs are more diverse and widespread now than they have ever been because, by any scientific definition, all birds are dinosaurs.

DIVERSIFICATION

In the aftermath of the K-Pg extinction the surviving birds underwent a huge adaptive radiation. Studies of the bird family tree have suggested that the majority of modern bird families established themselves in the first 15 million years after the impact.

The result of this explosion of variety is the over 10,000 species we see in the world today. Whether any of these families had already diverged from each other within the Cretaceous is still a matter for debate, although it seems likely that some must have done so. The example often pointed to for this is *Vegavis*, a bird from around the time of the extinction event that is theorized to have been an early ancestor of the duck family. But as is always the case with such limited evidence, it is exceptionally difficult to state anything for certain.

EXTINCTION ICONS

One sad tradition the birds retained from their dinosaurian ancestors was a continued fame for extinction. Best known of these is the Dodo (*Raphus cucullatus*), a flightless member of the pigeon family eradicated by hunting in the 17th century. Other ground-dwelling giants, such as the elephant birds and moas, which were the largest dinosaurs since the Mesozoic Era, met a similar fate after encounters with humanity.

TERROR BIRDS

Several times in the family tree, birds are seen abandoning flight and evolving into large, fully terrestrial forms. Among them are the phorusrhacids, which could reach around 10 ft (3 m) in height. Their "Terror Bird" nickname comes from their formidable beaks, which were capable of slicing into the flesh of prey. Some earlier examples of giant terrestrial birds from more recently after the K-Pg extinction event, such as *Gastornis*, were once also thought to behave in this way, but biomechanical modeling of their jaws suggests a more herbivorous diet. Surprisingly, the closest living relatives of the giant *Gastornis* are ducks and geese.

↑ Able to reach speeds of 240 mph (386 km/h) while dive-bombing its prey, the modern

Peregrine Falcon (*Falco peregrinus*) is technically the fastest dinosaur ever to have existed.

DISPLAYS

S cent and sound are a great way to communicate in the animal
world, but visual cues can be just as important. We commonly see
this in the vibrant plumage of modern birds, and evidence suggests it
was present in dinosaurs, too.

BILLBOARD STRUCTURES

Many dinosaurs evolved structures with large surface areas, which
were likely used as massive display features. Examples are the huge
sail of *Spinosaurus*, the crest of *Dilophosaurus*, and also potentially the
back plates of *Stegosaurus*. Usually with only the bones to go on,
the exact function of these structures remains unconfirmed, but the
display theory is logical considering their size and relation to features
seen in animals today. The vascularization of much of the bone is also
evidence of their function. Being so well supplied with blood implies
they needed to provide oxygen to overlying tissue, which makes sense
if they were being used for display.

Strong colorization has been suggested for the frills of ceratopsians,
too. Physical competition is energetically expensive and carries a high
risk of injury, so animals are often seen to evolve alternative initial
ways of competing to avoid it. Frill displays would certainly be one
such solution.

→ Many dinosaurs
may have used colored
feathers and display
patches on their skin in
the same way that their
modern counterparts,
the birds, do.

BROW RIDGES

Some dinosaurs had more subtle structures with no clear functional use, and these were therefore likely purely for display. A notable example is the brow ridges above the eyes of many theropod dinosaurs (such as *Allosaurus* and *Giganotosaurus*). The famed bull-horns of *Carnotaurus* were probably only for display too, being far too small to serve any use in defending the animal or taking down prey.

FEATHER FANS

Quill attachment nodes on the arm bones of several dinosaurs suggest they had spreads of full feathers over their arms, even if they may have lacked plumage elsewhere. These feathers were far too small to be of any use in flight and also too limited and oddly placed to serve as insulation. It's therefore logical to assume that their main use was for display. Many birds today color the areas around their wings for this purpose—the vibrant orange underside of the Kea (*Nestor notabilis*) is a notable example. Many theropod species may have colored their arm feathers for similar displays.

CRESTED DINOSAURS

The crests of dinosaurs were not always simply visual features, as some had complex internal structures that hint at more functional usage. No group of dinosaurs is seen to have a greater variation in head crests than the hadrosaurs.

HADROSAURS

Very common herbivorous dinosaurs during the Cretaceous, hadrosaurs were known for a time as "duck-bills," due to the wide, flattened shape of some of their mouths. A successful group of animals, regularly the dominant herbivores in their environment, they could be found across the Northern Hemisphere as well as down into what is now South America, although they are most known from extensive collections in the United States and Canada. It is believed they spent much of their lives walking on two legs until they reached sizes where this might become more difficult, at which time their forelimbs could work effectively as legs in a quadrupedal stance.

CREST VARIETY

Beside that of the *Parasaurolophus*, other famed hadrosaur crests include the forward-facing, rectangular structure on the skull of *Lambeosaurus* and the semicircular ridge of *Corythosaurus*. Both these hadrosaurs lived in North America during the Late Cretaceous and both were large herbivores, measuring about 50 ft (15 m) and 30 ft (9 m) long, respectively. These crests also had complex internal structures, creating S-shaped passages that could have been used to produce low-frequency calls, possibly for attracting mates or competing. Such sounds would have been picked up easily by these hadrosaurs because they had specialized, rod-like ear bones that would be particularly sensitive to these calls.

→ One of the most recognizable and bizarre of all dinosaur skulls, the hollow structure of the *Parasaurolophus* crest allowed it to be used for complex vocalizations.

~ *Parasaurolophus* ~

Of all the hadrosaurs, the head shape of one in particular has achieved iconic status. *Parasaurolophus* is famed for the enormous ridge of bone extending from the back of the skull and bending back around to form a curved crest. It has been suggested that the crest could have supported a membrane of skin, acting as a display sail, but no evidence of this has been found. The tubular, U-shaped structure of the crest, which may have allowed air to flow through it, and its connection to the nasal bones of the dinosaur, have led to suggestions that *Parasaurolophus* may have used it as a resonating chamber to amplify special calls. Finds of juvenile *Parasaurolophus* have shown that these crests began developing relatively early in life, having begun growth when the animal was only a quarter of its adult size.

SOFT-TISSUE CRESTS

The skull of another hadrosaur, *Edmontosaurus*, does not show any large, bony crest and it was for a long time assumed not to have had one. This changed in 2014 when a beautifully preserved specimen from Alberta, in Canada, was found with the mummified remains of a soft-tissue comb, similar to that of roosters, on the back of the head.

Hadrosaurs possessed particularly large eye sockets, indicating that vision was an important sense for them. This implies that, as with modern combed animals, the structure was likely brightly colored and a result of sexual selection. It raises the intriguing possibility that other hadrosaurs had similar soft-tissue structures that have not been preserved.

PARENTING

T he closest relatives of dinosaurs, the birds, are particularly attentive parents, insulating, guarding, and feeding their brood around the clock until they are ready to fledge the nest, and often even afterward. Speculating on the parental behavior of dinosaurs from fossil evidence is difficult, but plenty of clues remain.

EGGS

Like crocodiles and birds, dinosaurs for the most part laid hard-shelled eggs made of calcium carbonate. These eggs were composed of two layers, through which pores allowed for gas exchange. Uniquely among dinosaurs, the eggs of many theropods possessed three layers, which is also seen in modern birds. It is believed that this may indicate where dinosaurs switched from fully burying their eggs to partial burial and incubation by a brooding adult.

It has recently been claimed that some earlier dinosaurs had softer eggs, comparable to those of modern turtles. If correct, this would mean that hard shells evolved multiple times in reptiles, rather than just the once.

Shelled eggs are very energetically expensive for an animal to produce, and so even the largest dinosaurs laid relatively small eggs. Even the largest sauropods laid eggs only about the size of a soccer ball, so tiny in comparison to the gigantic multi-ton adults.

~ Embryos ~

In some exceptional cases the remains of unhatched dinosaur embryos have been found preserved inside their eggs. These juveniles show several features associated with modern animals, such as a small egg-tooth on the tip of the snout to help them break through the shell (seen in some titanosaurs). Scanning technology has helped reveal how far through their incubation some of these embryos may have been. The presence of extra teeth in embryos of the early sauropod *Massospondylus* are an example of this, as they would have been reabsorbed or lost before hatching.

OVIRAPTORS

The Cretaceous theropods known as oviraptors are so named because they were first discovered in association with a fossil nest in Mongolia in 1923. At the time it was assumed the animal found with the nest was preying on the eggs, using its sharp beak to penetrate the shells. For this reason oviraptors were called "egg thieves." In the years since, this has been revealed as a misnomer; far from predating another dinosaur's eggs, the eggs actually belonged to oviraptorid dinosaurs. Some of the most caring parents of all dinosaurs, remarkable fossils have shown oviraptors brooding over clutches of eggs much like modern birds, using their feathers to insulate them. Buried by sudden landslips, the parent dinosaur was preserved in the exact position in which it guarded the eggs in life.

← Although very rare, remarkable finds have shown dinosaur embryos still preserved inside their eggs. Sometimes their contents are known only by scanning them.

GROWING UP

Even the largest dinosaurs started off small. A baby titanosaur might have only been about 12 in (30 cm) high at the waist on first hatching, and yet would grow into an absolute giant. As dinosaurs grew, they laid down new layers of fibrolamellar bone (a woven bone matrix found in fast-growing animals). Growing quicker in more bountiful seasons, this growth was staggered and could produce growth rings, allowing some dinosaur bones to be dated in cross section, almost like a tree. More detailed histological studies of the bone structure can reveal more about the growth process.

Smaller dinosaurs had slower growth rates—the 2¹/₂-ft (75-cm) ornithopod *Jeholosaurus*, for example, is thought to have reached full size in about two years. In contrast, sauropods had the quickest growth rate, with *Apatosaurus* thought able to increase by 5¹/₂ US tons (5,000 kg) per year, reaching a full length of 75 ft (23 m) in just 15 years.

Dinosaurs lived to a range of ages, some theropods reaching a maximum of about 30 years, whereas the large sauropods may even have achieved a century.

→ Almost like counting tree rings, periods of fast and slow growth in some dinosaur bone cross sections can help us estimate their ages.

↘ Studies of juvenile *Jeholosaurus* have provided an insight into its growth rate, with the most rapid growth occurring in the first two years.

↓ *Apatosaurus* was one of the largest dinosaurs of Jurassic North America. The remains of many adults and juveniles have been found in the famous Morrison Formation, allowing paleontologists to study how this giant animal grew.

FLORAL INTERACTIONS

During the dinosaur age, Earth saw revolutions in the plant world, none of which was more impactful than the origination of the angiosperms, the flowering plants. Herbivorous dinosaurs evolved multiple ways to tackle the various plants of the time.

GASTROLITHS

Almost all grazing mammals today are capable of rotary chewing, but this was not the case for dinosaurs. This was a particular problem for the tough fern plants on which many of the dinosaurs fed. Needing assistance to break down the fibrous material, many dinosaurs turned to eating rocks. By swallowing small pebbles along with their plant food, the dinosaurs were able to physically grind food inside their stomachs. After some time processing, these stones were then passed, having been highly polished by the mechanical action and stomach acids. The stones, known as gastroliths, are regularly found with the remains of many dinosaurs today, especially those of sauropods.

CHEWING

In order to effectively chew food, dinosaurs may have had to evolve cheeks, to prevent the food from simply falling out the sides of their mouths, and specialized teeth. Many dinosaurs collected their teeth in concentrated rows, called dental batteries, which could grind up food when mashed together. In some species these batteries could contain over 300 teeth each.

Multiple groups of ornithischians, including the ceratopsians, evolved keratin-sheathed beaks at the front of their mouths, with the teeth behind, which allowed them to effectively snip off plant material before grinding it.

GRAZERS

Alongside angiosperms, the Mesozoic also saw the evolution of the first grasses. For a long time it was thought that dinosaurs never interacted with any grass plants. However, in 2005 coprolites (fossilized dung), which almost certainly came from a dinosaur in Cretaceous India, revealed an assemblage of phytoliths (microscopic mineral grains contained within grass fibers) preserved inside. Although grass did exist at this time, it didn't dominate vast plains of land in the way it does today. And so, while some species of dinosaurs may have fed on grass, none specialized as full grazers in the same way as many mammal lineages.

↑ As the Mesozoic progressed, the plant landscape would have shifted to become more familiar to us today.

Large cycads and tree ferns were still dominant, although angiosperms and grasses had started to spread.

MAMMALS

The dinosaurs may have dominated the Mesozoic, but other groups were never far away. Mammals had been around since the Triassic Period too, and the commonly quoted idea is that they survived by staying small and hiding away from the dinosaurs. But there is some evidence to suggest that they played more than just a passive role in the dinosaur age.

One staggering fossil found in 2012 shows a mammal carnivore known as *Repenomamus giganticus* actively predating on a weakened *Psittacosaurus*, both preserved in the act by a sudden volcanic debris flow. It's an incredible find that shows mammals could not only take themselves off the menu, but also put the dinosaurs on theirs. However, this seems to be a rare exception. For the most part, the mammals of the Mesozoic remained small, the biggest so far known being comparable in size to a badger. But they made up for their small size by being adaptable, smart, and patient. Their time would come.

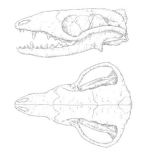

← Studies on the skull of *Repenomamus* have shown that it had a carnivorous diet. They also reveal how the complex mammalian inner ear bones evolved from bones that were originally associated with the jaws in their reptilian ancestors.

↙ *Repenomamus giganticus* is the largest known mammal from the Mesozoic Era, measuring 3 ft (1 m) from snout to tail. Living in the forests of Asia during the Early Cretaceous, it is one of the few mammals that could have been an active predator of dinosaurs (albeit only small juveniles).

CHANGING WORLD

By the end of the Cretaceous Period, the Earth's climate looked to be beginning another period of major change and upheaval. As with previous such events, much of the reason for this climatic shift could be traced to volcanic activity.

THE DECCAN TRAPS

In western India, a vast expanse of igneous rock evidences the presence of an enormous lava field 66 Mya. The flow deposits here cover a staggering 200,000 square miles (500,000 square kilometers), roughly the size of Spain. It is estimated that this area was active for 800,000 years. The geological feature left behind is known as the Deccan Traps.

Much like the Siberian Traps at the end of the Permian (252 Mya), aside from just destroying life in the local area, the huge quantities of sulphur dioxide (SO_2) and other gases from this continuous field of eruption would have altered the climate. Initial cooling from the SO_2 and from ash blocking sunlight was followed by a warming period of around 7.2°F (4°C), due to the increased carbon dioxide (CO_2) acting as a greenhouse gas.

These eruptions continued for some time after the end of the Cretaceous, too, at much the same rate as they had before, making it even harder for Earth to recover.

~ On the brink of extinction? ~

Whether or not the dinosaurs were truly in danger of extinction at the end of the Cretaceous, there can be no doubt that life was in a precarious position. Given time to recover from the climatic turbulence, there is nothing to suggest the dinosaurs couldn't have reigned for another 100 million years. However, one key intervention would change all that (see page 110).

DINOSAUR DECLINE?

Having peaked in diversity during the Cretaceous Period, diversity studies of the fossil record suggest that the dinosaurs may have been on the decline as a whole from as early as 73 Mya. The destabilization of the dinosaur food chain may have made them more prone to a sudden collapse if hit by a sudden environmental change. This idea remains somewhat controversial, as patterns such as this are often hard to distinguish from bias in the fossil record. The availability of access to fossil sites, volumes of collection, and the likelihood of preservation all make this a constant issue in paleontology.

↑ Today, the Deccan Traps, once a huge lava field, are visible as a very large province of igneous rock that sculpts a dramatic landscape across much of central and western India.

THE LAST DINOSAURS

There is a tendency to lump all dinosaurs together as coexisting at the same time. But this is, of course, incorrect, as dinosaurs spanned a period of over 160 million years. This length of time is best illustrated by the often-quoted fact that *Stegosaurus* and *Tyrannosaurus rex* are further apart from each other in time than *T. rex* is to us now. The question this then raises is: Which of the dinosaurs were actually present to witness the end of the Mesozoic?

The most famous names of that final stage of the Cretaceous (a time called the Maastrichtian) include *T. rex*, *Triceratops*, and *Ankylosaurus*. Living relatively close to the impact site, those species would also have felt some of its most violent initial effects. The hadrosaurs were among the most numerous and widespread groups at the time, possibly the final global success story of the non-avian dinosaurs. The giant sauropods were rarer, although they still persevered to the very end.

↓ *Endowed with a heavily armored skull,* Ankylosaurus magniventris *could reach almost 26 ft (8 m) in length. This formidable* herbivore lived in North America at the very end of the dinosaur age and was discovered in Montana's Hell Creek Formation.

→ The most iconic of the horned dinosaurs, *Triceratops horridus* only existed on the planet for the final 2 million years of the dinosaur era. The name comes from the three impressive facial horns, which could be used in competition with other members of the same species and also for fending off predators.

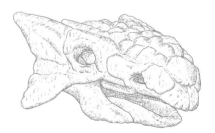

THE K-PG IMPACT

Although vulnerable, the dinosaurs did not seem to be at immediate risk of extinction at the end of the Cretaceous Period. This would all change in quite literally a single moment when, 66 Mya, just off the coast of what is now Mexico, Earth was struck by a gigantic asteroid.

IMPACT

It is estimated the rock that impacted the surface was up to 7½ miles (12 km) in diameter, hitting the surface at an approximate speed of 12 miles per second (43,200 mph/69,500 km/h). Geological analysis suggests that the asteroid was largely carbon-based, similar to the many objects that make up the asteroid belt in the solar system. It was not, as some previously claimed, an ice-based comet.

IMMEDIATE EFFECTS

Unlike the grind of other mass extinction events, many of the effects of an extraterrestrial strike are immediate, happening literally within seconds of the impact. The bedrock underlying the asteroid would have been vaporized, throwing a blast front in all directions. There would have been no escape for any animals within miles of the impact.

The seismic activity triggered by the asteroid sent shockwaves and mega-tsunamis around the globe. It's thought that some of the waves generated by the impact would have been 300 ft (90 m) in height, sweeping unstoppably over low-lying land.

← Asteroids are large rocks that orbit the Sun. Although common in our solar system, impacts on the scale of the Cretaceous event are exceptionally rare.

ASTEROID SEASON

Incredibly, we have evidence of the exact time of year the asteroid impact occurred. Tanis in North Dakota, in the United States, is an extraordinary site that appears to preserve the actual day of the impact. Remains of the victims reveals evidence of violent deaths, including dinosaurs swept away by tsunamis and fish choking on the glassy spherules ejected by the impact. Resident fish had reliable annual growth patterns recorded in their skeletons, faster at some times of the year, slow in others. Isotope analysis of these remains shows that these fish were killed during late spring, in around June. Combined with other evidence from the site, we therefore have a fairly reliable idea of when the asteroid hit.

Spherules of molten glass ejected due to the impact would have rained back down to Earth hundreds of miles away from the strike. These would have been hot enough to start major fires wherever they landed, with whole ecosystems burning uncontrollably for days.

LONG-TERM EFFECTS

The ash and debris cloud thrown up by the asteroid stayed in the atmosphere for years following the impact. So much material was ejected that the layer it formed blocked sunlight from reaching the Earth's surface, possibly for over a decade.

Without the light needed to photosynthesize, plants began to die on an unprecedented scale. The domino effect from this was a collapse of the food chain, as the herbivorous dinosaurs died away. Briefly, there was a glut of food for the carnivores, able to feed on the bodies of the many dead plant-eaters, but it was to be very short-lived, and they, too, would soon fall.

The impact marked the end of the Mesozoic Era and the start of our current era, the Cenozoic. More specifically, it divides the Cretaceous Period (K) from the Paleogene Epoch (Pg), hence the name of the event: the K-Pg extinction.

FINDING THE CRATER

Considering the scale of the catastrophe created by the asteroid hit, a natural assumption was that the impact crater would be obvious to find. However, with 66 million years for erosion and natural processes to cover it, this proved surprisingly difficult. The location was first narrowed down by following deposits of iridium. This element is rare on Earth, but is commonly found associated with extraterrestrial sources like meteorites. The K-Pg impact spread a clear layer of black iridium across the planet and, quite simply, the layer is thicker closer to the source.

A geomagnetic survey for oil exploration in 1978 did the rest, detecting a ring 110 miles (177 km) in diameter half off the coast of the Yucatán Peninsula, in Mexico. Dating the rock confirmed it to be 66 million years old, a match for the end-Cretaceous extinction. The crater was named for the small town found within it, Chicxulub.

↓ The Chicxulub crater can be found on the Yucatán Peninsula of Mexico. Although too large and covered to be seen fully, the cenotes (open cave reservoirs) that have formed in the limestone around the edge of the crater give some impression of its huge scale.

→ Around the world the K-Pg extinction can be seen in a clear transition in the fossils found in the rock. Mass assemblages of ammonites such as this could only be found in Mesozoic rocks, disappearing entirely in any strata younger than 66 million years old.

OTHER VICTIMS

T he non-avian dinosaurs were certainly the most famous victims of the end-Cretaceous extinction, but they were by no means the only group to suffer and meet their demise when the asteroid hit the Earth 66 Mya—the reptiles and birds were affected, too.

REPTILES

Notably the flying pterosaurs disappear at this point, just as they had reached their most magnificent sizes. However, this was somewhat expected, as the evidence suggests they'd been declining in diversity since their peak in the Early Cretaceous. It's possible that they had already begun to succumb to the pressure of competition from the birds that would supplant them.

Of the Mesozoic marine reptiles, the ichthyosaurs were, in fact, already long extinct by the end of the Cretaceous, having been slowly declining since their peak in the Jurassic. So, too, were the pliosaurs,

although their long-necked relatives, the plesiosaurs, had survived up to 66 Mya, only sadly also to become victims of the extinction with the dinosaurs. The same is true of the mosasaurs, which disappear from the fossil record at this boundary.

BIRDS

Although birds did make it through the extinction, it should be noted that they did not do so unscathed. The enantiornithines, which had been the most successful bird group of the Mesozoic, were wiped out along with their non-avian relatives. So, too, were the swimming hesperornithiforms. All modern birds today fall within the group Avialae, the only clade of dinosaurs to survive the K-Pg event.

AMMONITES

As a result of the extinction, the oceans lost a group now famed the world over for being iconic fossils: the spiral-shelled ammonites. These cephalopods had first evolved a little over 400 Mya during the Devonian Period, reaching their peak in the Mesozoic. Thanks to their abundance and the resilience of their shells, they are among the most commonly found fossils from the Mesozoic today. The largest known of the ammonites, *Parapuzosia*, could reach a diameter of 6 ft (1.8 m), although most were significantly smaller.

Other cephalopods, the belemnites, with their bullet-shaped shells, most likely went extinct at the K-Pg boundary as well. There are claims that a limited number of species may have survived for up to 20 million years afterward, but this is as yet unconfirmed by confident fossil evidence.

← Mosasaurs may have been the top of the marine food chain during the Late Cretaceous, but this dominance actually left them very vulnerable to extinction from a global catastrophe.

RECOVERY

As with any extinction event, the Earth would take a long time to recover, but it would do so with an explosion of diversity. With the dinosaurs gone, all the niches they once filled were now available for occupation by other animal groups. The lack of pterosaurs meant birds were now unchallenged in the skies, while the small, adaptable mammals were poised to seize the terrestrial world, just as the dinosaurs had done in the Triassic.

SHIFTING DIVERSITY

One unexpected change seen after the K-Pg event is an apparent shift in diversity. For all of earth history up until this point, life had been more diverse in the oceans than on land. However, through the Cenozoic (from 66 Mya to today) we see the opposite. The reason for this is largely a continuation of the Angiosperm Terrestrial Revolution (see Chapter 2, page 33). As flowering plants continued to radiate, so, too, did the insects alongside them. Insects are today the most diverse group of animals on the planet, with over 350,000 species of beetle alone, and this is largely because of the niches and complex ecological relationships they formed with plants as they evolved.

MAMMAL RADIATION

There are some examples of birds evolving to exclusively terrestrial lifestyles—such as the famed "Terror Birds" (see Chapter 7, page 93)—but for the most part it was the mammals that would claim the land. By the Cretaceous, the three major mammal clades (monotremes, marsupials, and eutherians) had already been established, but it was only after the K-Pg event that they radiated enormously in both diversity and disparity.

In some ways, mammals would evolve to match what came before them. The giant *Paraceratherium* was the sauropod of its time. At a little over 16 ft (5 m) in height, this rhino-relative is the largest known terrestrial mammal.

MEGAFAUNA

The idea that the age of mammals followed on from the dinosaurs is relatively well known, but the timescale is often misinterpreted. The most famous extinct mammals include the Woolly Mammoth (*Mammuthus primigenius*), Woolly Rhinoceros (*Coelodonta antiquitatis*), and Saber-Toothed Cats (*Smilodon*). However, these are all relatively recent mammals, existing only about ten thousand years ago and coexisting with humans. These animals, collectively referred to as megafauna, lived during the Last Glacial Maximum, a time colloquially known as the "Ice Age." Much like the dinosaurs, they are icons of extinct prehistoric life, but they are separated from the dinosaurs by the same 66 million years as ourselves. In the time between, a lot of fascinating fossil mammals were often overlooked.

~ Marine mammals ~

It was in the sea that the mammals outdid any Mesozoic reptile. Around 45 Mya, a group of mammals began evolving to a more aquatic lifestyle. Within 10 million years they had evolved to the group we now know as whales. Today the Blue Whale (*Balaenoptera musculus*) is not only the biggest mammal of all time, but also the biggest animal known, far larger than any dinosaur.

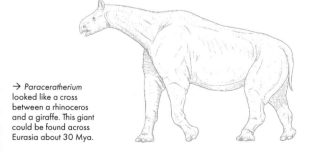

→ *Paraceratherium* looked like a cross between a rhinoceros and a giraffe. This giant could be found across Eurasia about 30 Mya.

NICHE EXPLORERS

T he dinosaurs evolved to exploit every niche available to them in their environments. Beyond the most obvious ones, such as ground-dwelling fern consumer and large terrestrial predator, they could also be found fulfilling more unexpected roles such as excavators and specialist insectivores.

BURROWING

Digging underground is a behavior usually associated with small animals, and there is evidence of this among dinosaurs. The American *Oryctodromeus* and Chinese *Changmiania* are examples that have been found preserved within their burrows. Both were Cretaceous ornithopods; they were bipedal with specialized forelimbs for tunneling. These arms were short but strong, with robustly built shoulders, primed for effectively removing dirt.

The burrows are preserved as infilled sandstone, distinctly different to the composition of the surrounding rock, and show that the burrows were not much bigger than the adult dinosaurs. Bones of juveniles found within the *Oryctodromeus* burrow along with the those of the adult show the dinosaur was using it as a sheltered hiding place for its offspring.

ANTEATERS

Mononykus was part of a dinosaur group called alvarezsaurids, characterized by long legs and arms reduced to the point that for a long time they were considered useless vestigial organs. *Mononykus* was unique in that it retained an individual large and curved claw on each arm. Range of motion work on its forelimbs have shown they would have been very effective at ripping back material, being immensely strong despite their small size. Combined with the shape of the claws it has been hypothesized that their primary function was to break into insect nests, in a similar way to modern anteaters.

NOCTURNAL HUNTERS

Not all dinosaurs would have been diurnal; some would have exploited the opportunities available at night. Although this behavior cannot be entirely confirmed, clues for nocturnal adaptations can be seen in the fossils. The most obvious are large orbits in the skull, which are indicators of enhanced sight in low light. Another is advanced hearing, as suggested by the structure of the ear canal. Both features can be seen in the Mongolian alvarezsaurid, *Shuvuuia*, making it a likely candidate for a nocturnal dinosaur.

↓ Barn Owls (*Tyto alba*) are modern dinosaurs adapted for nocturnal hunting, having facial feathers that aid in channeling sound. It is possible Mesozoic dinosaurs would have evolved similar features.

COLORS

When reconstructing dinosaurs their fossil skeletons can only tell us so much. One aspect that remained a mystery was their original color. The reason dinosaurs are so commonly depicted in greens and browns is purely because that is the typical palette of living reptiles and was not based on any hard evidence.

THE BREAKTHROUGH

The idea that we would never know the color of a dinosaur was a sad but accepted fact for the first 160 years of their study, but this would all change with a major discovery. The animal at the center of this was *Sinosauropteryx*, a dinosaur already famed for being the first discovered with fossilized feathers.

When looking at the feathers under a scanning electron microscope it was revealed that the smallest details had been preserved, including structures within the feathers known as melanosomes, only around 500 nanometers in size. Melanosomes can be found in modern bird feathers as well, and they are the structures that give the feathers pigment. Roughly speaking, different-shaped melanosomes correspond to different colors (more spherical being black, sausage-shaped reddish brown, and so on). What is true of the birds today was also true of the dinosaurs. By looking at the melanosomes in the fossil and measuring their abundance, aspect ratios, and organization, the corresponding color of the living animal could be interpreted.

Thus, in 2010, the first official dinosaur color was revealed. *Sinosauropteryx* was largely ginger in color, with white banding on its tail, possibly for use in signaling like a Ring-tail Lemur (*Lemur catta*). A later study would also reveal a racoon-like "bandit mask" patterning across its face, likely to protect its eyes from solar glare when hunting.

This technique has been used for several different dinosaurs to interpret their color. It has also been used for other prehistoric groups, including the ichthyosaurs. These marine reptiles showed clear countershading (dark on top, light on the bottom), which would have provided perfect camouflage in the sea and which is still seen in many ocean predators like the Great White Shark (*Carcharodon carcharias*).

~ Limitations ~

Although an incredible tool, the applications of a scanning electron microscope to determine dinosaur colors are limited, since it can only be used for specimens with exceptional preservation. It is still the case that for the vast majority of dinosaurs, only bone fragments remain, making this technique impossible.

However, as the vast extent of feather presence in dinosaurs is now known, birds can be more logically used when making inferences in reconstructions of plumage. Although the colors are usually still guesswork, this allows for a wider range of possible colors than the dull spectrum the group was previously given.

→ Reconstructed color patterns of *Sinosauropteryx* based on evidence from fossil melanosomes.

SOUNDS

When considering the sounds dinosaurs could have made, they are usually depicted as roaring. For some species, like the crested hadrosaurs, intricate structures in their skulls provide evidence for vocalizations, but for the vast majority, any detail on how they sounded was lost with the decay of their soft tissues.

The *Allosaurus* relative, *Aerosteon*, possessed a highly pneumatized (filled with air cavities) furcula (wishbone), which could suggest the presence of bird-like vocal chords (due to a clavicular air sac). However, this feature is absent from most other theropods, which, sadly, means roars were unlikely. It is more probable that they produced low grunts and coos, communicating in low frequencies like giant pigeons rather than bellowing roars. Many may even have been mostly mute, which is not unheard of in large animals today, such as rhinoceroses and giraffes.

So far only one confirmed fossilized syrinx (the avian equivalent of a larynx) has been discovered in a Mesozoic avian-dinosaur, *Vegavis*. However, perhaps in time more will be found that can shed light on dinosaur vocalizations.

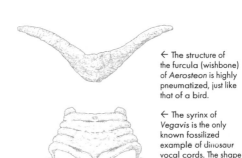

← The structure of the furcula (wishbone) of *Aerosteon* is highly pneumatized, just like that of a bird.

← The syrinx of *Vegavis* is the only known fossilized example of dinosaur vocal cords. The shape indicates it made duck-like honking and grunting noises.

← *Aerosteon riocoloradensis* was a theropod dinosaur from the Late Cretaceous of South America. Its name literally means "air bone," in reference to the bird-like structure of its bones, filled with air sacs, providing evidence for a similar respiratory system as birds.

SOCIAL CREATURES

As more is understood about dinosaurs through ongoing research, we can begin reimagining the interactions that made up their lives. Clues to their social behaviors, such as interspecies mixing, can be inferred from their modern relatives or seen in direct evidence from the fossil record.

SEXUAL DIMORPHISM

One of the biggest challenges to inferring behavior in dinosaurs is a lack of understanding of their sexual dimorphism. A common feature in animals today, it is certain many dinosaur species would have displayed this, but it is a remarkably hard thing to prove with fossil skeletons. There are plenty of suggestions for possible cases, from the plates of stegosaurs to the crests of hadrosaurs. But none of these has been proved beyond doubt. Without living specimens to examine, it's difficult to judge whether variations in individual fossils are due to the animals being different species, different sexes, or even at different stages of life (ontogeny).

INTERSPECIES MIXING

The remains of multiple herbivorous species of dinosaur have been found together in sites all across the world, although questions remain as to whether they had any real social interaction. In the wild today, animals are often seen traveling alongside herds of a completely unrelated species, usually for safety.

One trackway in Alaska provides a possible example, showing a herd of hadrosaurs traveling with several therizinosaurs, although this does throw up challenges, since we cannot know for certain how much they may have interacted. It could be that these were sought-out arrangements, benefiting both species, or perhaps merely a quirk of the environment forcing them together.

DISPLAY STAGES

As their closest relatives, birds are often employed to infer possible behaviors in dinosaurs. One such behavior seen in multiple species is the use of a display stage in attracting mates, where males will clear areas of land to perform in to attract females. Tentative evidence from Colorado, in the United States, in the form of concentrated scratch marks preserved in the ground, suggests this behavior may have been present in theropods. No definitive body fossils have been found with the marks, so the exact identity of the "floor scraper" is unknown.

↓ Fossilized trackways like these in Denver, Colorado, often show multiple different dinosaurs traveling together, although how much the species truly interacted with each other can usually only be guessed at.

BATTLE SCARS

Fossil bones hold reminders that each of these animals were living beings with individual stories to tell. Often this is shown in evidence of violent events: injuries sustained in their lives that may have caused death. The marks of such injuries are referred to as pathologies.

BROKEN BONES

The easiest pathologies to spot are usually broken bones that have healed. When the bone breaks, a hematoma forms around the fracture. Bone then grows to mend the break, although it doesn't do so perfectly. Differences in bone density can be picked up on scans, and several fractures can result in misshapen globules of bone rather than the original smooth surface.

Some individual dinosaurs were riddled with such injuries. One *Tenontosaurus* sustained three broken ribs and breaks to the hand and foot. It's possible the dinosaur survived an attack, but, unable to feed effectively as a result, died a few weeks later.

SIGNS OF DISEASE

Bones can even record diseases, giving an insight into the general health of some dinosaurs. The changes can take the form of extreme bone growth, as seen in the leg of one *Centrosaurus* specimen, which allow us to say that the unfortunate animal suffered from bone cancer. Bone growth from an osteomyelitis infection is also famously seen in the toe of the Wyoming *Allosaurus*, "Big Al."

Disease can present as a loss of bone too. Trichomonosis is caused by a parasite and presents as lesions on the jawbones of some theropods. Still seen in birds today, this disease can prevent the animal from feeding, leading to death through starvation.

PREDATOR-PREY INTERACTIONS

When carnivores feed on their prey they leave traces behind. This can be as simple as a series of cuts and gouges in the bones of the victims. If preserved in the fossil, measuring the size and shape of these marks and comparing them to the teeth of known predators in the area means it is possible to build up a picture of the food web for the ecosystem of the time. It can also provide evidence for the behavior of the predators—if the wounds have healed, it means the animal escaped, suggesting it was attacked while alive and not simply scavenged after death. Such healed wounds in *Triceratops* helped settle the debate as to whether *Tyrannosaurus rex* was an active hunter.

The strength required to leave observed marks on the bone can be used as data to infer an animal's bite strength. Feeding habits can also be inferred from the pattern of the scars—for example, whether the predator used puncture-pull head movement to strip flesh.

COMPETITION

Scars are often left by interactions between members of the same species. Raked blows across the frills of ceratopsians can be inferred to have come from the horns of other individuals, evidence that they locked together to compete, in a similar way to rutting animals today. The dome skulls of pachycephalosaurs can show trauma from repeated heavy impacts, suggesting they competed in this way, too.

ACCIDENTAL INJURY

Dinosaurs have been found with seemingly random injuries as well. These include one *Parasaurolophus* whose vertebrae and ribs record an impact theorized to have been caused by unfortunately getting caught underneath a falling tree.

→ Trichomonosis is a disease that causes lesions in the jaw (seen here as dark circles), and can be found in both Mesozoic theropods and modern birds.

FUTURE DIRECTIONS

Dinosaur paleontology is still very much in the middle of a golden era of discovery. Just 20 years ago, there were many unknowns that we thought impossible to determine, only for advancements in technology or stunning new finds to open up possibilities hitherto undreamed of. We can, however, rule out the idea of bringing dinosaurs back from extinction. Movies may have perpetuated the myth of this being achievable by extracting ancient DNA for cloning, but this is, sadly, impossible, since DNA cannot survive any form of the fossilization process.

So far, about a thousand species of dinosaur have been formally described and named, and it's estimated that about 30 new species are announced in scientific literature each year. With each new find comes the potential to discover more and resolve previously poorly understood relationships. One of the greatest aspects of paleontology is that these discoveries can be made by anyone, and come at any time.

← Fossils preserved in amber can show spectacular details of animals frozen in life, but any DNA extraction remains a sci-fi fantasy.

↙ The impressive claw of *Tyrannomimus fukuiensis,* an ornithomimid dinosaur that was first discovered and named during the writing of this book.

→ The dinosaur *Leallynasaura* was adapted for life in the cold southern continents, and is notable for its exceptionally long tail, which lacked ossified tendons. It is possible the dinosaur used this tail for signaling or even to wrap around itself to stay warm.

MUSEUM SENSATIONS

The only way to truly experience and understand the magnificence of dinosaurs is to see their fossils. This is why museums are so crucial in the story of these incredible animals. The first mounted dinosaur skeleton was "Haddy the Hadrosaurus," which was erected in the Academy of Natural Sciences, in Philadelphia, in 1868. An instant sensation, dinosaurs were soon a "must-have" exhibition for museums the world over.

To this day, museum specimens are sources of inspiration and wonder, connecting the science of paleontology with the public. Individual mounts can become tourist destinations in themselves, icons of their host cities. The biggest names include "Sue" the *Tyrannosaurus rex* of the Chicago Field Museum, the *Giraffatitan* of the *Museum für Naturkunde* in Berlin (currently the tallest dinosaur mount in the world), and "Dippy the Diplodocus," which greeted visitors to the Natural History Museum in London for over a century before going on tour around the United Kingdom.

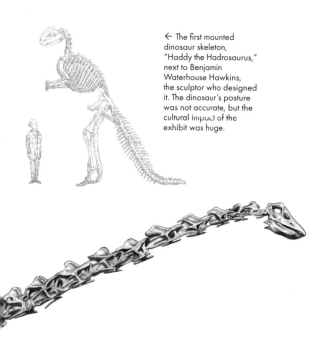

← The first mounted
dinosaur skeleton,
"Haddy the Hadrosaurus,"
next to Benjamin
Waterhouse Hawkins,
the sculptor who designed
it. The dinosaur's posture
was not accurate, but the
cultural impact of the
exhibit was huge.

← "Dippy the Diplodocus"
is one of the most famous
mounted dinosaur
skeletons in the world,
despite not containing a
single real fossil. The cast
is a composite of several
actual specimens, and
was donated to London
by Andrew Carnegie
at the request of King
Edward VII.

FLESH ON BONE

Impressive though dinosaur skeletons undoubtedly are, one of the most fascinating things about dinosaurs is imagining how they may have looked with muscle, fat, and skin on those bones. Recreating a creature no human has ever seen presents a unique challenge for artists and has resulted in a spectacular diversity of different styles.

CLASSIC DESIGNS

By the late 1890s the more recognizable look of a dinosaur had arrived. Although now highly outdated, the "tripod" pose given to many of the animals became iconic. Much of this work was pioneered by the American paleoartist Charles Knight, who created the blueprint for what was the quintessential dinosaur look for years to come. If anything, Knight was too successful, since his images were so vivid and popular that for many they remain the mental image conjured up by the word "dinosaur," even as science has moved past many of his speculative choices. Even those who have never heard his name are unknowingly influenced by the images he created.

Although many depictions are now inaccurate, Knight's work was regularly far beyond his time. His 1897 piece, *Leaping Laelaps*, shows two theropods fighting in extremely active poses, over half a century before the Dinosaur Renaissance (see Chapter 3, page 44) brought this way of thinking to the forefront of paleontology.

MONSTROUS SCULPTURES

With such limited evidence to go on and the field so young, it's no surprise that a lot of early artistic attempts were so wildly off how we now think dinosaurs looked. Perhaps the most famous examples are the inaccurate but charming full-size models in London's Crystal Palace Park. Built in the 1850s, the agile bipedal *Megalosaurus* was depicted as a stocky, quadrupedal, lizard-like giant, its long tail dragging across the ground. *Iguanodon* is also depicted more like a crocodile-elephant cross, with its thumb spike placed on the tip of its snout like a small rhinoceros. The statues, which still stand today, are a physical record of some of the first efforts at dinosaur reconstruction.

MOVIE MONSTERS

As the popularity of dinosaurs exploded at the turn of the 20th century, another medium gaining traction was the world of cinema. Naturally, it wouldn't take long for the two to mix and, in 1914, the first onscreen dinosaur was brought back from extinction, albeit in animated cartoon form, as "Gertie the Dinosaurus." But the playful antics of a cartoon sauropod would not be the archetype of dinosaurs in cinema for long. Already making appearances as the villains of written fiction, it was the giant theropods that were destined for the lead roles.

STOP-MOTION HEROES

For decades the best way to bring dinosaurs to life on film was through stop-motion—sculpting clay models and painstakingly moving them frame by frame to create the illusion of movement. The dinosaurs of Willis O'Brien helped make some of the most celebrated movies of the era: *The Lost World* in 1925 and *King Kong* in 1933. The most legendary name in the stop-motion world, Ray Harryhausen, even dabbled in dinosaur work. The models may now be outdated, having odd stances and overly lizard-like looks, but the dinosaurs of *One Million BC* (1966) and *The Valley of Gwangi* (1969) are Hollywood icons.

← The star of an early 12-minute movie by Winsor McCay, "Gertie the Dinosaurus" is one of the very first animated characters of any kind.

BOX OFFICE GIANTS

Unquestionably, the biggest name in dinosaur cinema is *Jurassic Park*. The 1993 box office smash put paleontology well and truly into the mainstream zeitgeist. For four years this dinosaur-based feature would hold the title of highest grossing movie of all time.

Yet more than just being successful financially, the movie also changed cinema—and paleontology—forever. Originally scoped as another stop-motion project, a mixture of incredible practical robotics and pioneering work in computer generated imagery (CGI) brought the dinosaurs to life in a way nobody had seen before. The dinosaurs in the movie were actually the first living animals ever animated with CGI (somewhat ironically, considering that they're extinct).

Though much of the movie is film fantasy, there can be no denying the effect the series had on science as well as popular culture. In the years following its release there was a significant uptick in people studying paleontology and evolutionary biology. The way it inspired young minds has legitimately improved the science, and that is surely worth the price of a hundred much more critically divisive sequels.

SCIENCE VERSUS FANTASY

Paleontology is very lucky to have ambassadors as popular and beloved as the dinosaurs, and science would be far worse off without their celebrity status. But there is often a price to be paid for this when the beauty and awe of the real animals are lost in a world of misconceptions.

MISUNDERSTOOD

In the years since *Jurassic Park*, dinosaurs have become regular stars, with CGI making it easier than ever to do what was once seen as impossible: to bring them into our world. But this isn't necessarily always good for the dinosaurs. They are now so often cast as the villains on screen that it is easy to forget they were real animals, with any natural behaviors or scientific accuracy lost in the inevitable over-the-top characterizations that come from being movie monsters.

TOYS IN RESEARCH

Due to the prevalence of dinosaur toys on the market, it could easily be claimed that they are more popular today than ever before. One of the great joys of dinosaur toy popularity is that they provide a visual reference for the progress of research. First surging in the 1950s, they capture a tangible scale of ideas, evolving from the lumbering, tail-dragging originals to modern forms incorporating feathers and more avian features. By the time someone reaches adulthood, their childhood dinosaur toys would be barely recognizable in the dinosaur toy aisles today.

Although it may sound trivial, studies on the production, spread, and popularity of dinosaur toys have been used to measure the dissemination of modern research, as well as in human-behavior experiments on perceived gender norms. They have proved a surprisingly insightful resource for study.

↑ A 1950s theropod toy showing the classic tripod pose and bloated body of a slow and lumbering animal, its arms in the incorrect "bunny hands" position.

↑ This 1970s sauropod toy has an oddly curvy spine and sprawling legs, which gives the animal an almost snake-like appearance.

↙ Although most are exceptionally inaccurate, there are likely more toys of *Tyrannosaurus rex* than any other animal on the planet, living or dead. The real animal had a much more horizontal stance, as opposed to the upright pose that is often seen in toys.

OTHER MEDIA

D evelopments in animation have helped dinosaurs spread beyond big budget movies and into all other forms of media. This has allowed them to dominate the cultural zeitgeist, with many dinosaurs now more globally recognizable than a lot of living creatures.

DOCUMENTARIES

Thankfully, the CGI revolution has meant dinosaurs can be brought to life more vividly in documentaries. Although others came before it, 1999's *Walking With Dinosaurs* is largely credited with showing the potential of styling paleontological documentaries in this fashion. Rather than just showing static fossils on the screen, dinosaurs could be seen as living animals, and part of vast prehistoric environments. Documentaries not only help inspire others to explore the science but also represent some of the most scientifically up-to-date and accurate representations of dinosaurs. As with all records of how the science has progressed, they may one day date as poorly as London's Crystal Palace statues (see page 133), but that will always be the way in a field such as paleontology.

VIDEO GAMES

Alongside movies and TV, dinosaurs were quick to spread to video games, too. First appearing in screen-elements of 1980s pinball machines, they quickly assumed the role of antagonists in some of the biggest games in the virtual world, such as *Tomb Raider*, *Mario*, and *Sonic*.

Sadly, for the most part, their roles in video games only accentuate their exaggerated reputation as monsters although, in return, software developed for these games has been used in multiple biomechanical studies reconstructing ancient modes of travel, especially when it comes to modeling flight. The code has been made to model objects obeying physical laws in the game, and this can be applied with surprisingly few alterations to modeling prehistoric creatures.

UNEXPECTED ADVANCES

Animation can also take the credit for some scientific advances. Scientific theory can only reveal so much and some ideas need to be put into practice. Animating dinosaurs can reveal issues and solutions not necessarily obvious during the initial scientific study. This artistic influence can be traced as far as back as *Fantasia* in the 1940s, which includes a sequence in which dinosaurs are shown moving out of their traditional upright pose and running in a similar way to how we would depict them today. The artists went against the then scientific knowledge simply because it didn't look natural when they tried animating it, and years later that instinct would be proved entirely right. Other interpretations in *Fantasia*, such as *Tyrannosaurus* having three fingers and living alongside the Jurassic *Stegosaurus*, would not hold up as well.

ART AND SCIENCE

I n modern science, paleontologists work closely with artists to reconstruct their findings accurately. The artists are an essential part of the scientific progress. No matter how much science can reveal about dinosaurs, there will always be unknown elements that cannot be reconstructed faithfully. The fossil record is simply too limited to preserve every detail. This is where reconstructions need to get creative.

ARTISTIC LICENSE

Professional paleoartists recreating the unknown should not be misconstrued as guessing blindly. Features added will be logical assumptions based on characteristics and behaviors observed in modern fauna. For example, in cases where the dinosaur's colors aren't known, educated assumptions can be made based on its habitat and by looking at living related animals in similar habitats. Features of camouflage and patterning are often repeated across various taxa.

INSPIRED BY MEDIA

There are plenty of examples of times when the norm has been reversed and pop culture has influenced science. For example, the tail spikes of stegosaurs are now widely referred to in scientific literature as "thagomizers," a term invented for a dinosaur joke in Gary Larson's *Far Side* comic strip in 1982.

Some paleontologists have also taken inspiration from pop culture when naming new species. There are plenty to choose from, although highlights include *Sauroniops*, a theropod with distinctive eye ridges named for the Eye of Sauron in Tolkien's works, and *Zuul*, an ankylosaur named after the demon from the 1984 movie *Ghostbusters* (with an additional species name of *crurivastator*, meaning "destroyer of shins").

Sometimes more elaborate features are added, including soft-body structures that may not lend themselves to fossilization and therefore stand little chance of being preserved. Such inferences fall into the realm of "speculative biology." They should be taken with a pinch of salt, but are an important part of the science, nonetheless. We see endless forms of beautiful diversity in today's animal kingdom, and there's no reason to think that dinosaurs weren't the same.

FOSSIL GEMS

One area in which paleontology and artistry have mixed in the past is in the making of jewelry and decorative pieces with the gemstone, amber. Formed from tree resin, amber has been known to trap and preserve animals, conserving them in exquisite detail. Although the animals most commonly seen are insects, there are rare instances of vertebrates being trapped and preserved in this way—and in one exceptional instance, part of a dinosaur. The fossil in question is small, only a 1¼-in (3-cm) section of the small dinosaur's tail, but all the detail is preserved, showing every filament on the feathers covering the thin structure. It's an exceptional find, but one marred in controversy due to ethical questions regarding mining amber specimens in Myanmar.

WORKING TOGETHER

Art has the ability to bring multiple aspects of dinosaur paleontology together. Paleontologists may work on individual facets of dinosaur biology, but it is only when that information is combined with the work of other specialist scientists, including paleobotanists and paleoclimatologists, that a full picture of the ancient world can be created by the artists.

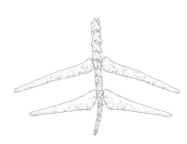

→ The four distinctive spikes at the end of the tail of *Stegosaurus* are now commonly referred to as thagomizers, although no word existed specifically for them until 1982.

PLANT PREDATORS

Large claws are regularly seen as being synonymous with "killer" animals. Claws can be devastating weapons when used in attack, and many theropods had them, such as on the hands of *Allosaurus* and the talons of *Velociraptor*. Therefore, when 3-ft (1-m) long claws were discovered in Mongolia in 1948, it was presumed they belonged to a large predatory theropod dinosaur (although they were initially described as belonging to a giant seaweed-eating turtle).

SCYTHE LIZARD

The claws did indeed come from a theropod, and this was dubbed *Therizinosaurus*, meaning "Scythe Lizard." This name is particularly apt since their primary function was likely for dealing with plant material. *Therizinosaurus* had made the evolutionary transition from carnivory to herbivory. Fossil material of *Therizinosaurus* itself is mostly limited to the arms and feet. However, just one genus in a family of animals—finds of related specimens like *Alxasaurus*—would reveal their body plan. The animal had a relatively stout body and short tail, stood bipedally, and its neck was quite long and slender with a comparatively reduced head size. At the end of the snout was a horned beak, perfect for slicing through foliage. Fossils from *Beipiaosaurus* have shown they were also covered in down-like feathers. *Therizinosaurus* could reach over 33 ft (10 m) in length and weigh as much as 5½ US tons (5,000 kg), making these herbivores some of the largest known theropods and the largest of all maniraptorans, a group usually associated with small, bird-like dinosaurs (and, of course, the birds themselves).

↘ Therizinosaurs were truly bizarre-looking animals, almost resembling a mash-up of other dinosaur elements. A successful group, they survived until the very end of the Mesozoic Era.

SAIL-BACK SAUROPOD

The necks of sauropods are already incredibly distinctive and over-engineered, but as is often the case in the massive diversity of life, there will be one species that takes things even further. In sauropods, this is the Argentinian *Amargasaurus*. Attached to the vertebrae of the neck were huge extensions of bone, some reaching up to 2 ft (60 cm) in length. Known as elongate hemispinious processes, it has been theorized that these structures supported a membrane of skin that stretched across them to form a display sail, possibly brightly colored, although the presence of such a membrane has not been confirmed.

Amargasaurus was conservative in size by sauropod standards, reaching about 30 ft (9 m) in length. Many of its mysteries remain hidden due to a lack of evidence, as currently only one skeleton is known of the animal, which was found in Argentina in 1984.

↓ The bizarre neck vertebrae of *Amargasaurus* show huge processes extending from the center of the bone. They were so large they would have limited the movement of the dinosaur's neck.

↓ The dorsal vertebrae on the back of the sauropod are also exceptionally weird, with large "paddle" processes that supported a raised structure along most of the dinosaur's back.

→ There is no scientific consensus on how the back of *Amargasaurus* would have appeared, or what its function may have been. The most common theory is that it supported tissue used for display, although others have suggested that its neck spines may have been fully exposed and covered in a keratin sheath for defense.

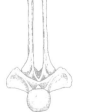

HUMPBACK CARNIVORES

In nature, optical display structures tend to be located in certain areas of the body. Head crests are common, as are wing and tail displays. *Concavenator*, however, is an anomaly, as it shows two hyper-extended vertebrae part way down its back, giving it a steep isolated hump. *Concavenator* was a theropod, related to *Carcharodontosaurus*, although for once the jaws of blade-like teeth might not be the first thing you'd notice. Forelimb bones of the animal suggest it was also at least partially feathered, so the hump may well have been covered with a brightly colored plumage, although its true function is still a matter of speculation.

Even more bizarrely, a completely unrelated and time-removed species, *Ichthyovenator*, shows the exact reverse of this. This species, like its relative *Spinosaurus*, had a sail along its back, but this was divided into two parts, with a triangular gap above the hip. Looking at them together they could almost have tessellated.

→ *Ichthyovenator* of Laos, in Southeast Asia, had the same skull and forelimb morphology as other spinosaurs but a very different dorsal ridge structure.

→ The skull structure of *Concavenator* shows its family connection with the much larger *Carcharodontosaurs* of South America.

← The Cretaceous theropod *Concavenator* is probably the most bizarre-looking dinosaur find to come from the famous Las Hoyas fossil site, which was found near the city of Cuenca in Spain. About the same height as an adult human, the bizarre hip structure marks it as unique among other theropods.

LIFE IN MINIATURE

Dinosaurs are typically associated with giant sizes, and while this was true for many of them, there were plenty of small dinosaurs to be found among them too. However, in some places there were no large dinosaurs whatsoever, and this led to the evolution of many strange surprises.

ISLAND LIFE

Due to higher sea levels in the Mesozoic Era, much of Europe was covered by a shallow sea, reducing the land area to a scattering of islands. While the islands could be teeming with life, they were systems highly limited in resources, making it impractical for any animals to sustain particularly large body sizes.

Where large animals struggle, their smaller counterparts thrive. Thus selection acts strongly for reduced body sizes, resulting in an evolutionary pattern known as "insular dwarfism." Essentially, an entire dinosaur ecosystem could survive here, but in miniature.

~ Dinosaurs of Hațeg ~

The best-known example of insular dwarfism in dinosaurs comes from what remains of Hațeg Island, now locked in mainland Romania. It preserves an ecosystem from the end of the Cretaceous, about 66 Mya.

Magyarosaurus were sauropods from Hațeg Island, smaller than most humans in height. This is still fairly large by overall animal standards, but it's nothing compared to the towering leviathans known elsewhere on the planet. Such sizes were matched by the hadrosaurs, the 16-ft (5-m) long *Telmatosaurus* being among the largest resident dinosaurs on the island.

Theropods made it to Hațeg as well, as seen in the bird-like *Balaur*. However, although enough of the skeleton is known to determine that this 6½-ft (2-m) long animal was definitely a theropod, no jaw remains have been found, making it hard to be certain whether this dinosaur was an endemic predator.

← A comparison between the femur of a Hațeg Island sauropod, *Magyarosaurus*, and that of a British sauropod, *Cetiosaurus*. The scale bar shows a length of 8 in (20 cm).

DEATH FROM ABOVE

With the reasonably land-locked dinosaurs so small, it was up to another group to fill the niche of top predator. On these Cretaceous islands this role was played by the azhdarchid pterosaurs, the largest animals ever to have flown. The wingspan of the largest, *Hatzegopteryx*, neared 40 ft (12 m) across, the size of a modern fighter jet. Even when landed they stood as tall as a giraffe. Towering over everything in the ecosystem, this was a world where pterosaurs would eat sauropods—an unimaginable scenario for their mainland relatives. Their 10-ft (3-m) long skulls atop long necks may have made them look unbalanced and incapable of flight, but they had hollow bones for weight reduction. That large head shifted their center of gravity forward, giving them much more control when launching quadrupedally from the ground.

ISOLATED MAMMALS

Insular dwarfism isn't just something that happened to dinosaurs either. It is a fact of evolution that can be seen repeatedly, even in groups familiar to us today. Less than a million years ago, islands in the Mediterranean were home to elephants only 39 in (1 m) in height. The *Palaeoloxodon* elephants of Sicily were one such example, interestingly alongside a species of giant mouse, the 26-in (65-cm) *Leithia*. Whereas the limited resources meant an evolutionary trend to small sizes for elephants, the lack of any large terrestrial predators had the inverse effect on some already small groups.

DINOSAUR BAT

T he skies of the Mesozoic were ruled by three groups of animals. The oldest were the arthropods (mostly insects), which were the most numerous and diverse fliers. Then there were the feathered avian-dinosaurs and the wing-fingered pterosaurs.

Mammals had yet to exploit the flying niche, since no bats are seen until after the Mesozoic (although their far-from-complete fossil record makes analyzing the precise timings of their evolution difficult).

UNIQUE FLIER

The maniraptoran dinosaurs that could fly all did so using the same method as modern birds: controlling air flow over a wing structure made up of multiple rows of strong airfoil feathers. However, one species was exceptionally different, *Yi qi*.

Dating from the Late Jurassic (160 Mya) of China, *Yi qi* is known from only one specimen found in 2007, but it is beautifully preserved. The body is clearly feathered and shows all the features that mark it out as a small theropod. But its arms are truly extraordinary. Running back from the wrist was a thin bone (styliform element), which was longer than the animal's ulna. Its function was to support a membrane of skin, bare of feathers, which started at the tip of its third finger. This created a wing unique among dinosaurs, appearing like a hybrid of a bat and bird. The tail is also odd—the typical dinosaur long, bony structure replaced by a short stub and long, emanating feathers.

Regularly considered a kind of "experiment of evolution," studies on its potential flight ability suggest that *Yi qi* was not particularly efficient. Easily outcompeted by the rise of birds, this extremely special dinosaur ended as a bit of an evolutionary dead end.

↘ The skeleton of *Yi qi* highlights just how bizarre this dinosaur was. Although from the outside, the wing looks superficially like that of a bat, it is not supported by modified fingers but rather by a separate bone element.

WEIRD AND WONDERFUL

The dinosaurs were an incredibly diverse and interesting group of animals and, in their over 160 million years on the planet, they evolved into a vast array of incredible forms. Some have been highlighted in this chapter, but there are endless more worth mentioning as well.

TOOTH VARIATION

Compared to mammals, dinosaurs on the whole had relatively simple tooth structures at first glance. Most showed homodont dentition, which means all the teeth were basically the same shape. The opposite of this is heterodonty, and it's a rare enough feature in dinosaurs that when one was found with heterodont teeth, it took the name *Heterodontosaurus*.

Mammals have heterodonty to a degree greater than any dinosaurs, but the heterodontosaurs came the closest. The very front of their mouths was covered by a keratin beak, a common feature among the ornithischians. Behind this were some pointed, tusk-like teeth, significantly larger than any other teeth in the jaw. At the back were rows of teeth similar to other herbivorous dinosaurs, indicating usage in grinding plant material.

← The skulls of *Heterodontosaurus* (top) and *Masiakasaurus* (bottom), two dinosaurs defined by their unique jaw structures and dentition. Both are from Africa but separated by 120 million years.

The function of the large tusks has been debated, with the main theories centered around usage in competition with each other, or in feeding. Although the back teeth clearly show adaptations for a herbivorous diet, the fact that the tusks are serrated has led some to suggest they could have been used in hunting, to supplement the dinosaurs' diet with occasional meat meals, in the same way that animals like pigs do today.

NO TEETH AT ALL: OVIRAPTOROSAURS

Many dinosaur species lost their teeth altogether. We see this in the ornithomimids, but also the oviraptorans. The largest oviraptorosaur of all, appropriately known as *Gigantoraptor*, stood 16 ft (5 m) high, making it taller than a *Tyrannosaurus rex*. Like its smaller relatives, it was incredibly bird-like in appearance. With a body covered in feathers, the head was held high at the end of a long, curved neck. Although the body may have made it look somewhat like a cassowary, the stout beak gave it an almost parrot-like face. Perfect for cutting through vegetation, this beak could also have been effective for meat too, or for whatever meal these generalist dinosaurs were tackling.

CURVED JAW

Masiakasaurus was a Madagascan theropod notable for its unique jaw structure. Like *Heterodontosaurus*, it had non-uniform teeth, although the weirdness doesn't end there. The jaw itself is the standout feature. At the tip of the snout the jaw appears to bend away from itself, the top curving up and the dentary (lower jawbone) curving downward. The curve on the lower jaw is so extreme that the first teeth point directly forward. The shape of the jaws and the robustness of the front teeth have been used as evidence that they were specialized for gripping prey, ensuring it couldn't escape.

ONGOING DISCOVERIES

This book covers a span across the vast dinosaur family tree, but this incredibly successful group of animals were so diverse and existed on the planet for so long that it is impossible to talk about them all. And with new species still being discovered in this new golden age of paleontology, there will always be more magnificent species to discuss.

GLOSSARY

adaptive radiation
Diversification of
organisms evolving
to fill niches of an
environment.

angiosperm
Plants that produce
flowers.

anoxia
A lack of dissolved
oxygen in water.

Archosauria
A group of reptiles
containing dinosaurs,
crocodiles, birds,
and pterosaurs.

arthropoda
Invertebrates with
exoskeletons and
jointed legs.

biodiversity
A measure of species
richness in a time or
environment.

biomechanics
The study of how animal
tissues move and react to
physical stresses.

caudal
Relating to the tail.

clade
A group of organisms
with a common
evolutionary ancestor.

cloaca
The single genital
opening seen in
crocodiles, birds,
and dinosaurs.

coprolite
Fossilized fecal matter.

dental batteries
Specialized tooth structures
evolved by dinosaurs for
processing plants.

diagenesis
The process of change as
sediment becomes rock.

fenestra
Structural openings
in the bone, usually in
reference to those in
the skull.

fibrolamellar bone
A transient bone tissue
found in fast-growing
animals.

gastrolith
A stone swallowed
by an animal to aid
in digestion.

Gondwana
Southern supercontinent
of South America,
Africa, Australia, and
Antarctica.

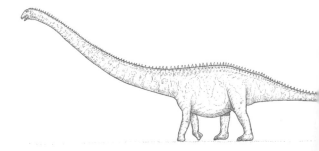

K-Pg extinction
The event in which the dinosaurs died out 66 Mya.

keratin sheath
A hard protein covering that forms the outer layer of claws and other structures.

***Lagerstätte* (pl. *Lagerstätten*)**
A site of exceptional fossil preservation.

Laurasia
Northern supercontinent of Europe, Asia, and North America.

megafauna
Very large animals, usually in reference to mammals of the Ice Age.

melanosomes
Microscopic structures in cells that store pigment.

Mesozoic
The geological era of time in which the dinosaurs lived (approximately 251–66 Mya).

microwear
Tiny patterns on the surface of fossil teeth that provide evidence of a dinosaur's diet.

morphology
The physical characteristics of an animal.

ornithischian
Dinosaur group with hip bones in a "bird arrangement."

osteoderm
Bony structure within the skin.

Pangaea
The supercontinent made of all major landmasses that existed when dinosaurs first evolved.

pathology
Physical evidence of injuries and disease.

pterosaurs
A flying reptile group from the Mesozoic.

saurischian
Dinosaur group with hip bones in a "lizard arrangement."

sexual dimorphism
Differences in physical appearance between different sexes of animal.

therapsids
Early relatives of mammals common before the dinosaurs.

ungual
The claw found at the end of an animal's finger or toe.

vestigial organ
A part of the body that has no apparent function, possibly a residual artefact of evolution.

wastebasket taxon
When fossils from multiple unique species have accidentally been classified as being only one.

FURTHER READING

Benton, M. J. 2019. *The Dinosaurs Rediscovered.* Thames & Hudson Limited, London.

Benton, M. J., and B. Nicholls. 2023. *Dinosaur Behaviour: An Illustrated Guide.* Princeton University Press, New Jersey.

Benton, M. J., and B. Nicholls. 2021. *Dinosaurs: New Visions Of A Lost World.* Thames & Hudson Limited, London.

Brusatte, S. 2018. *The Rise & Fall of the Dinosaurs.* Mariner Books, Boston.

Fastovsky, D. E., D. B. Weishampel, and J. Sibbick. 2021. *Dinosaurs: A Concise Natural History.* Cambridge University Press, Cambridge.

Naish, D., and P. M. Barrett. 2023. *Dinosaurs: How They Lived and Evolved.* Natural History Museum, London.

INDEX

ACKNOWLEDGMENTS

This book exists thanks to the excellent work and support of the publishing team, including Ruth Patrick, Lindsey Johns, Tugce Okay, and Ian Durneen, whose incredible artwork brought these animals back to life.

The process of researching the book was heavily aided by the fantastic team of the University of Bristol Palaeobiology group, of which I am lucky enough to be a member. In particular, I'd like to thank Dr. David Button, Dr. Nuria Melisa Morales García, Dr. Liz Martin-Silverstone, Dr. Ben Griffin, and Vicky Coules for their help in sourcing studies and answering the many random queries which arose while writing. Credit must also go to Dr. Tahlia Pollock, Elena-Marie Rogmann, Fionn Keeley, and Sophie Gayne who all contributed to making this book such a joyful project.

ABOUT THE AUTHOR

Rhys Charles is a paleontologist and science communicator. As head of the Bristol Dinosaur Project since 2016, he works with researchers to share the latest developments in paleontology with schools and the general public across the UK. He has also worked as a scientific consultant for multiple dinosaur books, video games, and large-scale events, as well as authoring original works, including *Frozen in Time: Fossils of the United Kingdom and Where to Find Them*. Rhys is based in Bristol, UK, but can often be found searching for fossils on the coast of his hometown of Penarth, Wales.